Sustainable Civil Infrastructures

Sustainable Civil Infrastructures (SUCI) is a series of peer-reviewed books and proceedings based on the best studies on emerging research from all fields related to sustainable infrastructures and aiming at improving our well-being and day-to-day lives. The infrastructures we are building today will shape our lives tomorrow. The complex and diverse nature of the impacts due to weather extremes on transportation and civil infrastructures can be seen in our roadways, bridges, and buildings. Extreme summer temperatures, droughts, flash floods, and rising numbers of freeze-thaw cycles pose challenges for civil infrastructure and can endanger public safety. We constantly hear how civil infrastructures need constant attention, preservation, and upgrading. Such improvements and developments would obviously benefit from our desired book series that provide sustainable engineering materials and designs. The economic impact is huge and much research has been conducted worldwide. The future holds many opportunities, not only for researchers in a given country, but also for the worldwide field engineers who apply and implement these technologies. We believe that no approach can succeed if it does not unite the efforts of various engineering disciplines from all over the world under one umbrella to offer a beacon of modern solutions to the global infrastructure. Experts from the various engineering disciplines around the globe will participate in this series, including: Geotechnical, Geological, Geoscience, Petroleum, Structural, Transportation, Bridge, Infrastructure, Energy, Architectural, Chemical and Materials, and other related Engineering disciplines.

SUCI series is now indexed in SCOPUS and EI Compendex.

More information about this series at http://www.springer.com/series/15140

Anand Tapase · Jeffrey Lee ·
Lei Zhang
Editors

Infrastructure Sustainability Through New Developments in Material, Design, Construction, Maintenance, and Testing of Pavements

Proceedings of the 6th GeoChina International Conference on Civil & Transportation Infrastructures: From Engineering to Smart & Green Life Cycle Solutions – Nanchang, China, 2021

Springer

Editors
Anand Tapase
Department of Civil Engineering
Karmaveer Bhaurao Patil College
of Engineering
Satara, Maharashtra, India

Jeffrey Lee
Department of Transport and Main Roads
Australian Road Research Board (ARRB)
Brisbane, QLD, Australia

Lei Zhang
Southeast University
Nanjing, China

ISSN 2366-3405 ISSN 2366-3413 (electronic)
Sustainable Civil Infrastructures
ISBN 978-3-030-79643-3 ISBN 978-3-030-79644-0 (eBook)
https://doi.org/10.1007/978-3-030-79644-0

This Springer imprint is published by the registered company Springer Nature Switzerland AG
The registered company address is: Gewerbestrasse 11, 6330 Cham, Switzerland

Introduction

This volume contains 14 papers that were accepted and presented at the 6th GeoChina 2021 International Conference on Civil & Transportation Infrastructures: From Engineering to Smart & Green Life Cycle Solutions in September 18–19, 2021, Nanchang, China. It contains research data, discussions, and conclusions focusing on a number of pavement materials and related geotechnical aspects of infrastructure. Topics include issues related to civil infrastructure such as use of construction waste, recycled aggregates, service life prediction of pavements, mechanical behavior of SMA, control measures of ready mixed concrete, determination of landslide high-risk areas, simulation of rock hydraulics in rock joint, and sustainable planning for provision of basic infrastructural facilities in rural areas of India. This information should lead to more resilient and sustainable infrastructure design, maintenance, and management. Various types of research were used in the various studies, including field measurements, numerical analyses, and laboratory measurements. It is anticipated that this volume will support decisions regarding the optimal management and maintenance of civil infrastructures to support a more resilient environment for infrastructure users.

Contents

About the Editors

Dr. Anand Tapase Assistant Professor at the Department of Civil Engineering of Rayat Shikshan Sanstha's Karmaveer Bhaurao Patil College of Engineering, Satara, is renowned Indian Civil Engineer with a research interest in pavement engineering, pavement materials, and sustainable constructions. He completed his Bachelor's degree in Civil Engineering from Rayat Shikshan Sanstha's, Karmaveer Bhaurao Patil College of Engineering, Satara of Shivaji University, Kolhapur, and both M. Tech and Ph.D. degrees from the College of Engineering, Pune (C.O.E.P.) of Savitribai Phule Pune University Maharashtra, India. He is working as Associate on an International Cooperative Research Project carried by Texas Transportation Institute, Texas A&M University, in collaboration with the Texas Department of Transportation, United States of America (USA).

He has two patents by his name. He is Editorial Board Member of International Journal, "Innovative Infrastructure Solutions, Springer Nature Switzerland AG". He is Associate Member of the American Society of Civil Engineers (ASCE) and Transportation and Development Institute (TDI). He is working as Consultant on various government projects. He has authored, co-authored, and edited refereed research 11 journal papers, 19 chapters (author/co-author/editor), and 36 conference papers. He is working as Reviewer for renowned international publications including International Journal of Pavement Research & Technology (IJPRT), Journal of Cleaner Production both published by Elsevier/Science Direct, Journal of Testing and Evaluation of ASTM International, Journal of Performance of Constructed Facilities of ASCE. He is been recognized as '*Outstanding Reviewer*' by the Journal of Cleaner Production published by *Elsevier/Science Direct*. Also, he has worked as Editor of a book volume for Springer Publication at GeoChina 2018. He was invited to serve as Session Chair/Moderator at the GeoChina 2018 International Conference in China. Presently, he is working as Member of the Technical Committee of International Conference GeoChina 2021 which will be held in Nanchang, China, in the year 2021, and working as Lead Editor for the Sustainable Civil Infrastructure Book Series, Springer.

Dr. Jeffrey Lee is Principal Pavement Engineer at the Australian Road Research Board (ARRB) responsible for the management and delivery of research projects for the Department of Transport and Main Roads and other members of the Australian Road Authorities. He has been providing advisory services to road authorities in the areas of asphalt pavement designs, non-destructive evaluation, and increase use of recycled material for pavement construction.

Jeffrey has been Peer Reviewer for different journals published by the American Society of Civil Engineers (ASCE) and American Society for Testing and Materials (ASTM). More recently, he is Chief Editor for an ASCE Geotechnical Special Publication (GSP) and Co-Chair of the Technical Organizing Committee for the upcoming GeoChina 2018 International Conference. He is current Queensland Branch President of the Australian Society for Concrete Pavements, Fellow of the Institute of Engineers Australia, Adjunct Professor at the Queensland University of Technology, and Adjunct Fellow at the University of Technology Sydney.

Prof. Lei Zhang is Professor of Intelligent Transportation System Research Center and Transportation School at Southeast University, China. He had worked earlier at California University, San Diego, as postdoc. His research interest covers non-destructive detection technology and pavement structure and material for transportation infrastructures including highway, steel deck bridge, airfield, and high-speed railway. He completed both his undergraduate and graduate studies at Southeast University. He has authored, co-authored, and edited more than 50 technical papers, 2 books, and 2 standards. He is Associate Editor of the International Journal of Pavement Research and Technology, and he is International Member of ASTM, Member of TRB (AFB65), and Co-Chair of WTC subcommittee. He was awarded as Outstanding Youth Teacher of Southeast University and was also included in Six Talent Peaks Project of Jiangsu Province.

Mechanical Behavior of SMA 8 Modified with Nano Hydrotalcite

João Crucho[1], José Neves[1(✉)], and André Pedro[2]

[1] CERIS, Department of Civil Engineering, Architecture and Georesources, Instituto Superior Técnico, Universidade de Lisboa, Lisbon, Portugal
{joao.crucho,jose.manuel.neves}@tecnico.ulisboa.pt
[2] Instituto Superior Técnico, Universidade de Lisboa, Lisbon, Portugal
andre.s.pedro@tecnico.ulisboa.pt

Abstract. Nanotechnology is currently being applied in pavement engineering and, in particular, asphalt mixtures. This emerging technology can enhance materials' properties and in recent years there has been research carried out on the influence of modifications through different nanomaterials. Hydrotalcite, which is a type of layered double hydroxide (LDH), has been regarded as a promising modifier, particularly on account of its ability to reflect UV radiation (improving UV aging resistance), delay oxygen permeability and retard the loss of volatile components. Such properties can actively contribute to enhance the aging resistance of the asphalt binder, and consequently of the asphalt mixtures. The objective of this paper is to present the effect of nano hydrotalcite modification in the mechanical behavior of Stone Mastic Asphalt with an 8 mm upper sieve size of the aggregate (SMA 8). The gap-graded aggregate skeleton bounded with a mastic mortar that is very rich in modified bitumen (7.5%), and contains 0.5% of natural cellulose fibers, are special characteristics that are able to induce a more significant effect of the nanomodification. The dosage of hydrotalcite was 3% by mass of asphalt binder. The study of the mechanical behavior comprised laboratory tests of resistance to permanent deformation (EN 12697-22), indirect tensile strength (EN 12697-23), water sensitivity (EN 12697-12), stiffness (EN 12697-26), and resistance to fatigue (EN 12697-24). The main conclusions were established by comparing the results obtained with those of an unmodified SMA 8. Although the unmodified SMA 8 already exhibited good mechanical performance, it was possible to observe that the nanomodified SMA 8 also showed an adequate behavior for application in surface layers. The most relevant enhancement was obtained in the evaluation of water sensitivity by the indirect tensile strength test, achieving a 100% indirect tensile strength ratio. In general, the results revealed the potential of hydrotalcite for the modification of asphalt mixtures, indicating positive effects in terms of durability.

1 Introduction

Technological innovations in pavement engineering are currently being explored to improve materials' performance from the mechanical, economic, and environmental points of view. In the case of surface layers, the asphalt mixtures have to withstand a

A. Tapase et al. (Eds.): GeoChina 2021, SUCI, pp. 1–10, 2021.
https://doi.org/10.1007/978-3-030-79644-0_1

wide range of environmental and traffic conditions and, at the same time, provide special and more exigent conditions of skid and rutting resistance, among other characteristics. In addition to this, bitumen is a material that is sensitive to aging, and its properties deteriorate over time also affecting the performance of asphalt mixtures (Sirin et al. 2018). Aging is a particularly important factor in surface layers because those pavement layers are exposed directly to environmental conditions such as ultraviolet (UV) radiation, moisture, oxygen, and greater temperature variations. New asphalt mixtures have been developed, such as Porous Asphalt (PA) and Stone Mastic Asphalt (SMA), using a wide variety of additives in order to enhance the performance of asphalt mixtures during the pavement's life cycle. In recent years, different additives were developed for different purposes: adhesion improvers, fibers, rubber, warm mix asphalt additives, rejuvenators, and polymers (Behnood 2019, 2020; Caputo et al. 2020; Gupta et al. 2019; Habbouche et al. 2020; Picado-Santos et al. 2020; Rossi et al. 2017; Slebi-Acevedo et al. 2019). These new asphalt materials not only provide more resistant and resilient, and longer life pavements but also protect the environment and reduce maintenance costs.

Self-healing is aimed at increasing the lifespan of road pavements by reducing the aging effect on asphalt pavements. Three main self-healing technologies are currently available for asphalt pavement construction and maintenance: nanoparticles, induction heating, and rejuvenation (Behnood 2019 Crucho et al. 2019; Pasupunuri et al. 2017; Tabaković and Schlangen 2016; Vo et al. 2020).

The application of nanomaterials as an additive in asphalt mixtures is being considered today as an option in pavement technology. Indeed, like other additives, nanomaterials are suitable for asphalt binder modification with a view to improving asphalt mixtures' performance. Several studies have demonstrated the benefits of the modifications with nanomaterials – such as nanosilica, nanoclays, and nanoiron – on the mechanical performance and aging resistance (Crucho 2018; Crucho et al. 2018, 2019; Filippi et al. 2018; Martinho and Farinha 2017; Yanga and Tigheb 2013).

Hydrotalcite is a type of layered double hydroxide (LDH) (Evans and Slade 2005; Wang and O'Hare 2012). LDHs are ionic lamellar compounds made up of positively charged brucite-like layers of organic-inorganic or inorganic-inorganic nanomaterials. Compared to clays mineralogical structure, LDHs are less prone to delamination due to stronger interactions interlayers. The substance has already been considered to be a promising modifier for construction materials, particularly on account of its ability to reflect UV radiation (improving UV aging resistance), delay oxygen permeability, and retard the loss of volatile components. There are research studies on potential applications of LDH in concrete materials with very positive results (Duan et al. 2013; Xu et al. 2009). The application of LDH in asphalt mixtures has also been studied in recent years mainly thanks to its ability to be a physical barrier against UV and, as a consequence thereof, to delay the aging of bitumen (Cui et al. 2016; Liu et al. 2014; Pang et al. 2014; Peng et al. 2015; Xu et al. 2015, 2016, 2017; Zhao et al. 2015; Zhu et al. 2019).

We believe that these properties of nano hydrotalcite can actively contribute to enhancing the aging resistance of an asphalt binder, and consequently of asphalt mixtures. The influence could be more effective in the case of binder-rich asphalt mixtures. The innovative objective of this paper is to present a study on the feasibility of the nano hydrotalcite modification of SMA 8. The paper describes a preliminary laboratory study on the mechanical behavior characterized by indirect tensile strength and water sensitivity, stiffness, and resistance to fatigue and permanent deformation.

2 Methodology

2.1 Materials

An SMA 8 asphalt mixture was used in this study. The aggregates mixture consisted of the following percentages (relative to the total mass of aggregates mixture) of the three fractions based on the minimum and maximum nominal sizes (d/D): 20% of 6/12 fraction; 54% of 2/6 fraction; and 19% of fraction 0/6. Figure 1 shows the grading curve of the aggregates mixture. Aggregates were obtained from crushed gabbro rock.

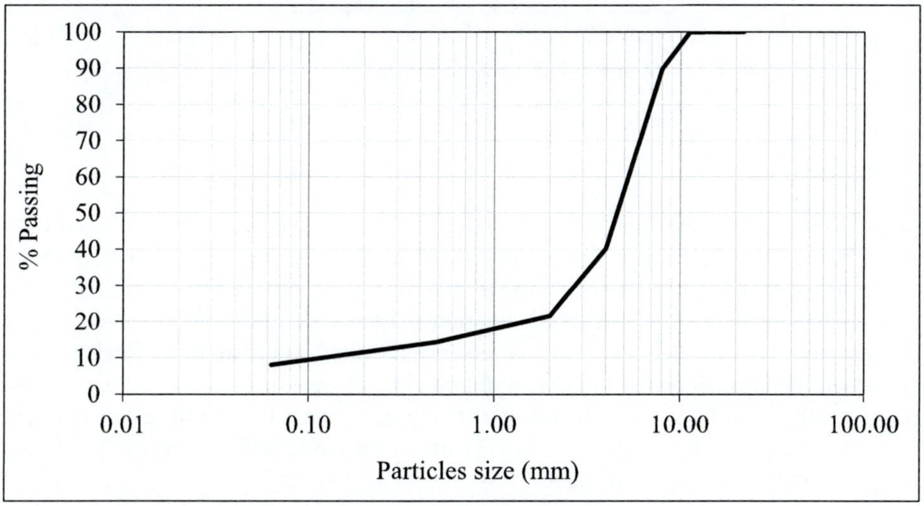

Fig. 1. Particles size distribution

The SMA was made up of 7% (relative to the total mass of aggregates mixture) of filler (limestone) and 0,5% (relative to the total mass of asphalt mixture) of natural cellulose fibers. This mixture was designed in accordance with EN 13108-5. The asphalt mixture was made up of 7.5% (relative to the total mass of asphalt mixture) of PMB 45/80-65 modified bitumen (control binder). Table 1 presents the main Marshall properties of the asphalt mixture.

The asphalt mixture was modified with 3% (relative to the total mass of asphalt binder) of nano hydrotalcite. Crucho (2018) describes the modification procedure followed in this study. The properties of penetration (EN 1426) and softening point (EN 1427) of the control and modified binders are presented in Table 2. Two similar asphalt mixtures were produced in the study: the control mixture (unmodified with hydrotalcite), and the modified mixture (modified with 3% of hydrotalcite).

Table 1. Marshall properties of control asphalt mixture

Property	Unit	Value
Stability	kN	6.7
Flow	mm	3.2
Marshall quotient	kN/mm	2.1
Porosity	%	4.9

Table 2. Properties of control and modified binders

Binder	Penetration	Softening point
	$\times 10^{-1}$ mm	°C
Control	52.5	90.0
Modified	47.1	93.0

2.2 Laboratory Tests

Table 3 lists the properties and test standards followed in the laboratory study according to the European specification framework in EN 12697 for hot mix asphalt. Indirect tensile strength and water sensitivity were determined according to EN 12697-23 and EN 12697-12, respectively. Stiffness evaluation was based on EN 12697-26 and adopting the four-point bending beam equipment, under controlled strain, at 20°C. Resistance to fatigue was determined using the four-point bending beam equipment, under controlled strain, at 20°C and with 10 Hz sinusoidal loading, in accordance with EN 12697-24. Evaluation of permanent deformation was based on EN 12697-22 (wheel-tracking), using the small-size device and specimens conditioned in air.

Table 3. Test methods

Property	Test standard	Parameters
Indirect tensile strength	EN 12697-23	Indirect tensile strength (ITS)
Water sensitivity	EN 12697-12	Indirect tensile strength ratio (ITSR)
Stiffness	EN 12697-26	Stiffness modulus Phase angle
Resistance to fatigue	EN 12697-24	Fatigue lines
Permanent deformation	EN 12697-22	Rut depth in air (RD_{air}) Proportional rut depth in air (PRD_{air}) Wheel-tracking slope in air (WTS_{air})

3 Results and Discussion

3.1 Indirect Tensile Strength and Water Sensitivity

The mixture modified with 3% hydrotalcite presented good performance in the indirect tensile strength (ITS) and indirect tensile strength ratio (ITSR). Compared with the control mixture, the modified mixture presented 3% increase in ITS_{dry} (under dry conditions, with no water sensitivity conditioning) and 9% increase in ITS_{wet} (after water sensitivity conditioning). The results of ITS and ITSR of the control and modified mixtures are presented in Fig. 2. The control mixture also presented a good performance for these parameters, indicating no problems with ITS or ITSR. This means that the margin for enhancement using the modification was limited. Nevertheless, in the modified mixture the increase in ITS_{wet} superior to the increase in ITS_{dry} led to an increase in ITSR, with the modified mixture presenting a ratio of 100%. Such results provide a good indication of the positive effect of the modification with hydrotalcite on ITS and water sensitivity.

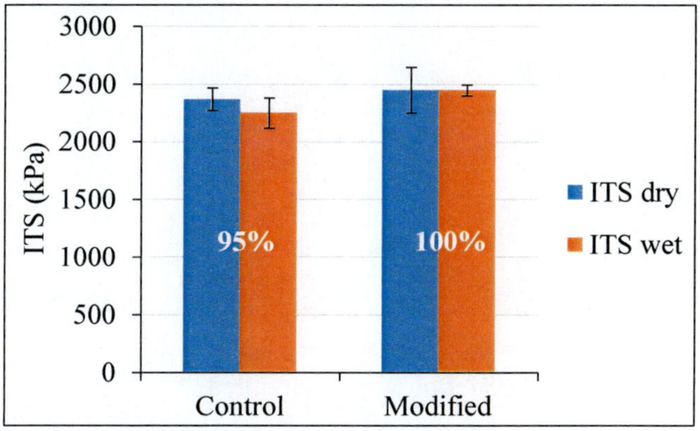

Fig. 2. ITS and ITSR of control and modified mixtures

3.2 Stiffness

The modification with hydrotalcite had some effect on the stiffness of the asphalt mixture. Compared with the control mixture, the modified mixture presented a slightly lower stiffness modulus, with a variation dependent on the loading frequency. The reduction in stiffness modulus of the modified mixture was 20% and 4% at 1 Hz and 30 Hz respectively, and for the intermediate frequencies, the observed reduction was approximately linear. With regard to phase angle, the modified mixture presented slightly higher values than the control mixture. At 1 Hz the increase in phase angle was 7% and at 30 Hz the increase in phase angle was 18%, meaning the values of the intermediate frequencies were within these two extremes. The results of the stiffness modulus and phase angle of the control and modified mixtures are presented in Fig. 3 and Fig. 4 respectively.

The elastic slopes (slope of the regression line of the natural logarithms of the loading frequency and elastic modulus) calculated from the stiffness data were 0.316 and 0.377

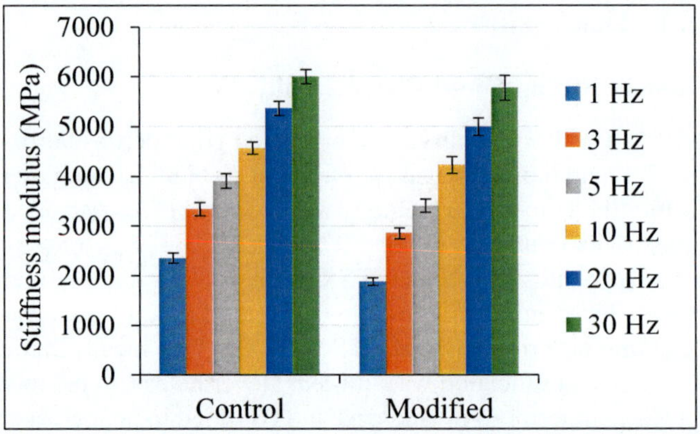

Fig. 3. Stiffness modulus of control and modified mixtures

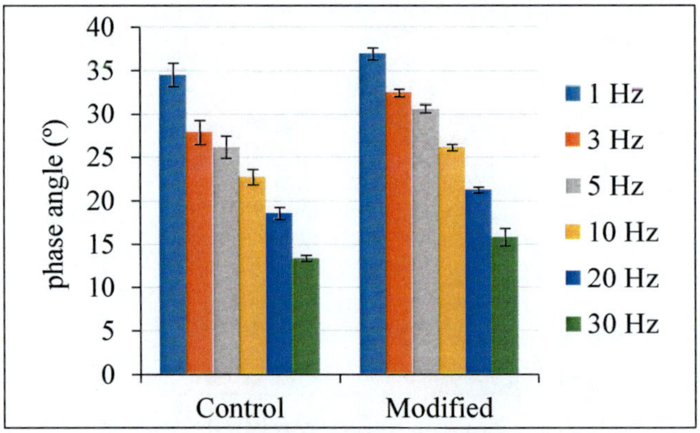

Fig. 4. Phase angle of control and modified mixtures

for the control and modified mixtures respectively. The elastic slope alone is not an accurate indicator of the aging level of the asphalt mixture; nevertheless, this parameter does decrease as the aging increases (Crucho 2018). As the modified binder presented a lower penetration value and higher softening point, one possible explanation for the higher elastic slope of the modified mixture, in line with a slightly lower stiffness modulus and slightly higher phase angle, could be that it is an effect of aging protection of the asphalt binder and mixture, particularly in the mixing and compaction stages. This property requires further investigation.

3.3 Resistance to Fatigue

The modification with hydrotalcite had an effect on fatigue resistance. However, the differences observed were relatively small considering the high dispersion commonly observed in fatigue tests. The fatigue lines of the control and modified mixtures are presented in Fig. 5. The strain corresponding to the application of one million loading cycles (ε_6) was 255 µm/m for the control mixture and 224 µm/m for the modified mixture. Thus, the modified mixture presented a reduction in ε_6 of 12%. The slopes of the fatigue lines were −5.0 and −5.7 for the control and modified mixtures respectively. In the case of the slope of the fatigue line, the modified mixture presented an increase of 13%.

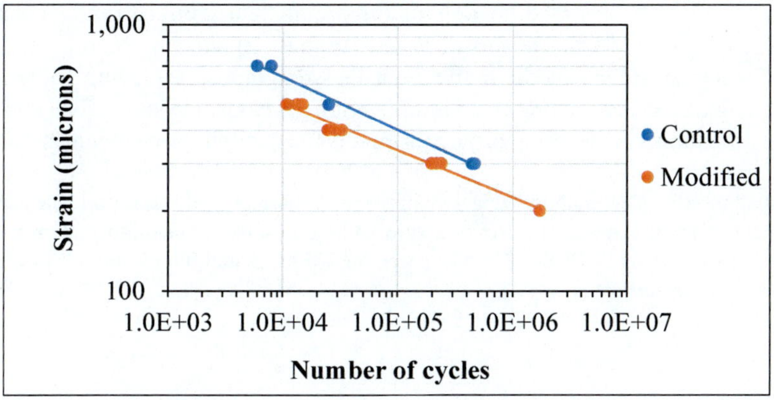

Fig. 5. Fatigue lines of control and modified mixtures

3.4 Permanent Deformation

The parameters obtained with the wheel-tracking tests, i.e. rut depth in air (RD_{air}), proportional rut depth in air (PRD_{air}), and wheel-tracking slope in air (WTS_{air}), for the control and modified mixtures are presented in Table 4. For each parameter, the respective value of standard deviation is presented in brackets. The modified mixture presented higher values for all parameters, indicating a softer behavior, which to some extent, is in line with the results for stiffness. Nevertheless, the results obtained can be considered within the range expected for SMA 8 mixtures.

Table 4. Wheel-tracking results of control and modified asphalt mixtures

Asphalt mixture	RD_{air}	PRD_{air}	WTS_{air}
	mm	%	mm/10^3 cycles
Control	3.9 (3.1)	7.5 (5.9)	0.286 (0.206)
Modified	10.9 (1.7)	21.4 (3.2)	0.343 (0.134)

4 Conclusions

This paper presents and discusses the results of a laboratory study aimed at assessing the effect of nano hydrotalcite modification on the mechanical behavior of an asphalt mixture. The selected mixture, a binder-rich (with 7.5% bitumen) Stone Mastic Asphalt with 8 mm upper sieve size of the aggregate (SMA 8), was modified with 3% hydrotalcite. The unmodified and the modified asphalt mixture were tested for indirect tensile strength, water sensitivity, stiffness, resistance to fatigue and permanent deformation.

The results obtained from the experimental study allow for the conclusion that nano modified SMA 8 presents in general an adequate mechanical performance for application in surface layers. The modification with hydrotalcite had effects on the mechanical performance of the SMA 8. The most important enhancement was obtained in the evaluation of water sensitivity by the indirect tensile strength test, achieving 100% indirect tensile strength ratio. The hydrotalcite modification gave positive indications regarding a potential effect in terms of aging protection and consequent improvements in durability.

Acknowledgments. The authors acknowledge the Laboratory of Transport Infrastructures (Department of Civil Engineering, Architecture and Georesources, Instituto Superior Técnico, University of Lisbon) and CERIS – Civil Engineering Research and Innovation for Sustainability – for the support of this study.

References

Behnood, A.: Application of rejuvenators to improve the rheological and mechanical properties of asphalt binders and mixtures: a review. J. Clean. Prod. **231**, 171–182 (2019). https://doi.org/10.1016/j.jclepro.2019.05.209

Behnood, A.: A review of the warm mix asphalt (WMA) technologies: effects on thermo-mechanical and rheological properties. J. Clean. Prod. **259** (2020). https://doi.org/10.1016/j.jclepro.2020.120817

Caputo, P., et al.: The role of additives in warm mix asphalt technology: an insight into their mechanisms of improving an emerging technology. Nanomaterials **10**, 1202 (2020). https://doi.org/10.3390/nano10061202

Crucho, J.: Development of an accelerated asphalt concrete aging method and utilization of nano-modifiers to improve durability of asphalt concrete. Ph.D. Thesis, Instituto Superior Técnico, Universidade de Lisboa, Portugal (2018)

Crucho, J., et al.: Mechanical performance of asphalt concrete modified with nanoparticles: nanosilica, zero-valent iron and nanoclay. Constr. Build. Mater. **181**, 309–318 (2018). https://doi.org/10.1016/j.conbuildmat.2018.06.052

Crucho, J., et al.: A review of nanomaterials' effect on mechanical performance and aging of asphalt mixtures. Appl. Sci. **9**, 3657 (2019). https://doi.org/10.3390/app9183657

Cui, P.Q., et al.: Characteristics of using layered double hydroxides to reduce the VOCs from bituminous materials. Constr. Build. Mater. **123**, 69–77 (2016). https://doi.org/10.1016/j.conbuildmat.2016.06.117

Duan, P., et al.: Influence of layered double hydroxides on microstructure and carbonation resistance of sulphoaluminate cement concrete. Constr. Build. Mater. **48**, 601–609 (2013). https://doi.org/10.1016/j.conbuildmat.2013.07.049

Evans, D.G., Slade, R.C.T.: Structural aspects of layered double hydroxides. In: Duan, X., Evans, D.G. (eds.) Layered Double Hydroxides. Structure and Bonding, vol. 119, pp. 1–87. Springer, Heidelberg (2005). https://doi.org/10.1007/430_005

Filippi, S., et al.: Effect of nanoadditives on bitumen aging resistance: a critical review. J. Nanomater. **2018**, 2469307 (2018). https://doi.org/10.1155/2018/2469307

Gupta, A., Rodriguez-Hernandez, J., Castro-Fresno, D.: Incorporation of additives and fibers in porous asphalt mixtures: a review. Materials **12**, 3156 (2019). https://doi.org/10.3390/ma12193156

Habbouche, J., et al.: A critical review of high polymer-modified asphalt binders and mixtures. Int. J. Pavement Eng. **21**(6), 686–702 (2020). https://doi.org/10.1080/10298436.2018.1503273

Liu, X., et al.: Fatigue properties of layered double hydroxides modified asphalt and its mixture. Adv. Mater. Sci. Eng. **2014**, 868404 (2014). https://doi.org/10.1155/2014/868404

Martinho, F., Farinha, J.: An overview of the use of nanoclay modified bitumen in asphalt mixtures for enhanced flexible pavement performances. Road Mater. Pavement Des. (2017). https://doi.org/10.1080/14680629.2017.1408482

Pang, L., et al.: Effect of LDHs on the aging resistance of crumb rubber modified asphalt. Constr. Build. Mater. **67**, 239–243 (2014). https://doi.org/10.1016/j.conbuildmat.2013.10.040

Pasupunuri, S.K. et al.: Self-healing pavements: a revolution in pavement materials. In: 18th Crossroads World Road Meeting 2017, India (2017). https://www.researchgate.net/publication/327838600_SELF-HEALING_PAVEMENTS_A_REVOLUTION_IN_PAVEMENT_MATERIALS

Peng, C., et al.: Preparation and properties of a layered double hydroxide deicing additive for asphalt mixture. Cold Reg. Sci. Technol. **110**, 70–76 (2015). https://doi.org/10.1016/j.coldregions.2014.11.013

Picado-Santos, L., et al: Crumb rubber asphalt mixtures: a literature review. Constrt. Build. Mater. **247**. https://doi.org/10.1016/j.conbuildmat.2020.11857

Rossi, C.O., et al.: Adhesion promoters in bituminous road materials: a review. Appl. Sci. **7**, 524 (2017). https://doi.org/10.3390/app7050524

Sirin, O., et al.: State of the art study on aging of asphalt mixtures and use of antioxidant additives. Adv. Civil Eng. **2018**, 3428961 (2018). https://doi.org/10.1155/2018/3428961

Slebi-Acevedo, C.J., et al.: Mechanical performance of fibers in hot mix asphalt: a review. Constr. Build. Mater. **200**, 756–769 (2019). https://doi.org/10.1016/j.conbuildmat.2018.12.171

Tabaković, A., Schlangen, E.: Self-healing technology for asphalt pavements. In: Hager, M.D., van der Zwaag, S., Schubert, U.S. (eds.) Self-Healing Materials. APS, vol. 273, pp. 285–306. Springer, Cham (2016). https://doi.org/10.1007/12_2015_335

Yanga, J., Tigheb, S.: A review of advances of nanotechnology in asphalt mixtures. In: 13th COTA International Conference of Transportation Professionals (CICTP 2013). Procedia - Social and Behavioral Sciences, vol. 96, pp. 1269–1276. https://doi.org/10.1016/j.sbspro.2013.08.144

Xu, S., et al.: Facile preparation of pure CaAl-layered double hydroxides and their application as a hardening accelerator in concrete. Chem. Eng. J. **155**(3), 881–885 (2009). https://doi.org/10.1016/j.cej.2009.08.003

Xu, S., et al.: Investigation of the ultraviolet aging resistance of organic layered double hydroxides modified bitumen. Constr. Build. Mater. **96**, 127–134 (2015). https://doi.org/10.1016/j.conbuildmat.2015.08.019

Xu, S., Yu, J., Zhang, C., Yao, T., Sun, Y.: Effect of salicylic acid intercalated layered double hydroxides on ultraviolet aging properties of bitumen. Mater. Struct. **49**(4), 1235–1244 (2015). https://doi.org/10.1617/s11527-015-0573-y

Xu, S., et al.: Performance evaluation of asphalt containing layered double hydroxides with different zinc ratio in the host layer. Pet. Sci. Technol. **35**(2), 127–133 (2017). https://doi.org/10.1080/10916466.2016.1248774

Vo, H.V., et al.: Effects of asphalt types and aging on healing performance of asphalt mixtures using induction heating method. J. Traffic Transport. Eng. (English Edition) **7**(2), 227–236 (2020). https://doi.org/10.1016/j.jtte.2018.10.009

Wang, Q., O'Hare, D.: Recent advances in the synthesis and application of layered double hydroxide (LDH) nanosheets. Chem. Rev. **112**(7), 4124–4155 (2012). https://doi.org/10.1021/cr200434v

Zhao, Z.J., et al.: The aging resistance of asphalt containing a compound of LDHs and antioxidant. Pet. Sci. Technol. **33**(7), 787–793 (2015). https://doi.org/10.1080/10916466.2015.1014965

Zhu, K., et al.: Flame-retardant mechanism of layered double hydroxides in asphalt binder. Materials **12**, 801 (2019). https://doi.org/10.3390/ma12050801

Application of Graded Crushed Stone Base in Urban Road Asphalt Pavement

Peiliang Zhang[1,2](✉), Jinyan Liu[1,2], Yingbiao Wu[1,2], and Jinjin Shi[1,2]

[1] Cangzhou Municipal Engineering Company Limited, Cangzhou 061000, China
[2] Hebei Province Road Materials and Technology Engineering Technology Research Center, Cangzhou 061000, China

Abstract. Combining the characteristics of graded broken stone (GBS) and SRB, a pavement base structure with cement stabilized recycled masonry aggregate (CRMA) and upper base of GBS was adopted in a road project in Cangzhou City (CZC). The indoor test determined that the optimal water content (OWC) of the GBS mixture was 3.2% and the maximum dry density (MDD) was 2.197 g/cm^3; OWC of CRMA is 14.3%, and MDD is 1.81 g/cm^3. The investigation results of pavement deflection and cracks in the six years of opening to traffic show that the asphalt pavement structure with the lower base (LB) of SRB and the upper base (UB) of GBS has good crack resistance.

1 Preface

Most of the urban roads in my country are mainly SRB structure, which has the advantages of high overall strength, high carrying capacity, good water stability (WS), relatively low cost, and easy guarantee of basic performance and quality. However, the shrinkage cracks of SRB material itself are unavoidable, which will lead to pavement diseases such as RC and water damage. The road will be damaged after being put into use for 2 to 3 years, which is far from reaching its designed service life. This type of damage will affect the traffic service function and increase maintenance costs.

The use of a good graded gravel for UB and SRB for LB can largely prevent and reduce RC on SRB. This structural feature enables the graded gravel layer to fully absorb the strain energy (SE) released by the cracks in the lower layer, thereby achieving the effect of restraining the cracks [1, 2]. UB of GBS can play a role in isolation, improve the temperature and humidity conditions of LB of SRB, and reduce the temperature change, temperature gradient and humidity change of LB of SRB [3]; At the same time, CRMA is a SRB material with a small elastic modulus (EM), which reduces the temperature shrinkage and dry shrinkage of SRB, and is beneficial to reduce cracks.

Combined with the test section (TS) of LB of CRMA and UB of GBS in a road project in CZC, the mix ratio design (MRD) of GBS was carried out, and the key points of construction technology and quality control were summarized. Through the investigation of pavement deflection and cracks in the past 6 years, the anti-cracking effect of this structure is analyzed.

A. Tapase et al. (Eds.): GeoChina 2021, SUCI, pp. 11–20, 2021.
https://doi.org/10.1007/978-3-030-79644-0_2

2 Engineering Situation

A road project in CZC is a planned north-south branch road, bi directional and four lane (BDFL), and the total thickness of the road surface structure is 56 cm. General pavement structure (GPS): 3.5 cm thick AC-13F ordinary asphalt concrete (OAC) top layer + 4.5 cm thick AC-16C OAC middle surface + 18 cm thick cement stabilized gravel (CSG) (R_7: 3.0 – 4.0 MPa) +two steps 15 cm thick lime soil (LS) ($R_7 \geq 0.7$ MPa). Stake number K0 + 280–K0 +580 is TS, and stake number K0 + 020–K0 + 280 is general comparison sections (CS). The base layer of the pavement structure of the test section is designed as 18 cm GBS + 15 cm cement stabilized recycled masonry(CSRM) (R7 $\geq$ 2.0 MPa) + 15 cm LS (R7 $\geq$ 0.7 MPa). The road project was completed in June 2013.

3 Composition Design

3.1 Design of Aggregate Composition of CSRM

The brick-concrete construction waste is crushed and screened to form recycled masonry aggregates (RMA) with continuous gradation. The particle composition of RMA is shown in Table 1. The liquid limit W_L of fine-grained soil below 0.6 mm is 27.3%, the plastic limit W_P is 19.8%, and the plasticity index I_P is 7.5; The average crush value of RMA is 33.8%. Its particle composition complies with the industry standard "Code for Construction and Quality Acceptance of Urban Road Engineering" CJJ 1-2008 CJJ 1-2008 Cement Stabilized Aggregate Gradation for Sub-dry Road Base Course.

Table 1. Particle composition of RMA

Screen size /mm	37.5	31.5	26.5	19.0	9.50	4.75	2.36	1.18	0.60	0.075
Pass quality percentage /%	100	93.9	88.5	76.9	56.8	40.7	30.8	18.4	13.3	2.4
Gradation range	90 –100	-	66 –100	54 –100	39 –100	28 –84	20 –70	14 –57	8–47	0–30

The cement adopts CZC 'Tie Shi' brand P.S32.5 retarded cement, the initial setting time is 185 min, the final setting time is 288 min, the flexural strength (FS) of 3 days and 28 days are 3.8 MPa and 6.5 MPa, the compressive strength of 3 days and 28 days are 17.1 MPa and 38.6 MPa respectively, and each index meets the specification requirements.

The amount of cement in CSRM will directly affect the road performance of the mixture. The cement dosage of CSRM was tested in proportions of 4.5%, 5.0%, 5.5%, and the design results of the mix ratio are shown in Table 2. The test results show that using CSRM with a cement dosage of 5.0% as LB, OWC is 14.3% and MDD is 1.81 g/cm^3.

Table 2. Mix ratio test result

Mixture proportion (cement dosage)	OWC (%)	MDD (g/cm^3)	7d unconfined compressive strength (MPa)	$\overline{R_7} \geq R_d/(1\text{-}Za\cdot Cv)$ (MPa) ($R_d = 2.0$ MPa)
4.5%	13.6	1.80	$\overline{R_7} = 1.96$ $S = 0.240$ $C_V = 0.123$	$\overline{R_7} = 1.96 \leq 2.5$
5.0%	14.3	1.81	$\overline{R_7} = 3.4$ $S = 0.199$ $C_V = 0.059$	$\overline{R_7} = 3.4 \geq 2.2$
5.5%	15.1	1.82	$\overline{R_7} = 4.0$ $S = 0.205$ $C_V = 0.051$	$\overline{R_7} = 4.0 \geq 2.2$

Note: R_d——concrete structure;
C_V——Coefficient of variation of test results;
Za——The coefficient that varies with the guarantee rate (or confidence level α) in the standard normal distribution table is taken as the guarantee rate 95%, Za = 1.645.

3.2 Composition Design of GBS Mixture

The strength of GBSB is mainly derived from the strength of the crushed stone itself and the intercalation between the particles. Therefore, it is necessary to select high-quality crushed stone and carry out a reasonable composition design so that the grading of the crushed stone can form a skeleton-dense structure. This project adopts clean, non-weathered limestone aggregate with sufficient strength and good particle shape. The test results of aggregates with different particle sizes are shown in Tables 3 and 4.

Table 3. Coarse aggregate test results

Project	Needle flake particle content (%)			Crushing value (%)
	5 –10 mm	10 –20 mm	20 –30 mm	
Test results	9.5	12.2	11.9	16.4
Skills requirement	≤ 20			≤ 30

Table 4. Fine aggregate test results

Project	Apparent density (g/cm^3)	Bulk density (kg/m^3)	Porosity (%)	Particle content (<0.075 mm) (%)	W_L (%)	Plasticity index
Test results	2.469	1.453	41.2	12.8	22.9	3.1
Skills requirement	—	—	—	≤20	< 25	< 4

Table 5. GB of mixture

Screen Size (mm)	20–30		10–20		5–10		0–5		Actual SG (%)	Design Gradation	
	15.0%		30%		25%		30%				
	Original mineral	Percentage of mixture	Original mineral	Percentage of mixture	Original mineral	Percentage of mixture	Original mineral	Percentage of mixture		range	Median
37.5	100.0	15.0	100.0	30.0	100.0	25.0	100.0	30.0	100.0	100 –100	100.0
31.5	100.0	15.0	100.0	30.0	100.0	25.0	100.0	30.0	100.0	90 –100	95.0
19	13.9	2.1	80.4	24.1	100.0	25.0	100.0	30.0	81.2	73 –88	80.5
9.5	0.3	0.0	2.6	0.8	95.4	23.9	100.0	30.0	54.7	49 –69	59.0
4.75	0.1	0.0	0.0	0.0	7.0	1.8	98.4	29.5	31.3	29 –54	41.5
2.36	0.1	0.0	0.0	0.0	0.3	0.1	72.1	21.6	21.7	17 –37	27.0
0.6	0.1	0.0	0.0	0.0	0.3	0.1	39.8	12.0	12.1	8 –20	14.0
0.075	0.1	0.0	0.0	0.0	0.2	0.1	14.3	4.3	4.4	0 –7	3.5

According to the results of grading and screening of raw materials, the mixing ratio of GBS and the synthetic gradation (GB) are determined, as shown in Table 5. Through the heavy compaction test, it is determined that MDD of the mixture is 2.197 g/cm^3 and OMC is 3.2%.

4 Construction Process

After the lower bearing layer (LBL) is prepared, pave LB of CSRM. After the 7 days curing period is over, pave UB of GBS.

The GBS mixture does not contain any binder, and its strength mainly comes from the interlocking effect between the aggregates. Therefore, the construction of GBS is more difficult. In addition to considering the quality of raw materials, reasonable grading design, and optimal structural layer combination design. A scientific construction plan should be designed and strict construction quality control should be carried out. Only by paving GBSB into a high-quality structural layer with uniform, high density, and good water permeability can ensure the performance of its functions such as crack reduction, fatigue resistance, and drainage [4].

4.1 Mixing and Transportation

The centralized plant mixing method is used to produce CSRM mixture and GBS mixture. During production, the cement dosage of CSRM is 0.5% higher than the mix ratio. Due to the high water absorption of RMA, the moisture content of the mixture is 1 –2% higher than OWC; The moisture content of GBS is controlled within the optimal moisture content range of 0 –0.5%. In order to ensure that the gradation of the mixture is controlled within the standard range, the changes in the water content of the mixture should be closely monitored during production and adjusted in time.

In order to ensure the quality of the mixture, enclosed transport vehicles are used for transportation to prevent the mixture from evaporating too quickly; During the transportation, the vehicle should try to avoid bumps and sudden braking to prevent the mixture from separating.

4.2 Paving

The mixture is transported to the construction site by a dump truck and spread with a paver at a paving speed of 1.0 m/min –1.5 m/min with no pause in the middle to reduce horizontal joints. The loose paving coefficient of CSRM is 1.40, and the thickness of virtual paving is controlled to 21 cm; The loose paving coefficient of GBSB is 1.33, and the thickness of the virtual paving is 24 cm. A full-time quality inspector is set up on the construction site to control construction quality, inspect width, flatness, elevation and other items, as shown in Fig. 1.

The compaction effect is the best when CSRM and GBSB is equal to or slightly higher than OWC. In order to prevent the surface moisture from evaporating, the road roller will follow up tightly after the mixture is spread. In the paving and rolling process, the water content must be strictly controlled. When the water content is low, use a spray sprinkler for watering.

Fig. 1. Paving GBSB

4.3 Rolling

LB of CSRM rolling combination is: static pressure with 22T vibratory roller for 2 times and then vibratory compaction for 4 times; with 21T three-wheel roller for two times, check the degree of compaction after the track depth is not more than 5 mm. Stop rolling after the degree of compaction meets the design requirements. Strictly control the rolling time. Sprinkle a small amount of water on the surface to keep the surface moist during the regimen period, and avoid excessive sprinkling.

The strength of GBSB is mainly derived from the interlocking of coarse particles obtained by rolling and the bonding strength formed by the filling of fine aggregates. Therefore, improving the quality of rolling work is a direct means to improve the structural strength of graded crushed stone. The key to improving the quality of rolling lies in which rolling process is used to make the degree of compaction meet the requirements, without crushing the material, and ensuring the uniformity of strength [5–7]. Through the trial paving of the test section, the construction and rolling combination of GBSB of this project is: rolling with a 26T rubber roller for 2 times; rolling with a 22T single-drum roller for 3 times, including static pressure, weak vibration, and strong vibration; Use a 21T three-wheel roller to roll twice; use a 26T rubber roller to roll once, as shown in Figs. 2 and 3.

Fig. 2. Rolling of graded gravel base

Fig. 3. The effect of GBSB after rolling and forming

5 Quality Control and Long-term Monitoring

5.1 Compaction Detection

The construction quality was controlled on site, and the compaction detection of the LB of CSRM and UB of GBS was strengthened. The compaction of each layer was measured at 6 points by the sand filling method. The data is shown in Table 6.

Table 6. Compactness test result

Number	Point position	Compactness (%)	
		Cement stabilized recycled masonry lower base	Graded gravel upper base
1	East side of K0 +290	97	102
2	East side of K0 + 350	98	101
3	west side of K0 + 440	98	99
4	East side of K0 + 470	99	104
5	East side of K0 + 550	96	100
6	west side of K0 + 575	96	103

Layer of CSRM: average compaction degree $\overline{K} = 97.3\%$, standard deviation $S = 1.1\%$, representative value $K = 95.5\% > 95\%$, and single-point compaction degree $K_L \geq 95\%$, the pass rate is 100%.layer of GBS: $K = 101.5\%$, $S = 1.2\%$, $K = 99.5\% > 97\%$, and $K_L \geq 97\%$, the pass rate is 100%.

5.2 Deflection Detection

5.2.1 Base Deflection Detection

The top surface of GCSB in the test section and the top surface of the cement stabilized crushed stone base (CSCSB) in the comparative section were respectively tested for deflection. BDFL were all longitudinally set with a measuring point every 15 m. The data is shown in Table 7.

Table 7. Base deflection value comparison

Road section	Base	Average flexure index (0.01 mm)	Standard deviation	Represents flexure index (0.01 mm)
TS	GBS	28.8	8	42
CS	Cement stabilized macadam	17	6.3	27.4

The design flexure index (DFI) of GCSB is 57.8 (0.01 mm). According to the data in the above table, although the deflection value of GBSB is greater than that of CSCSB, it meets the design requirements.

5.2.2 Surface Flexure Index Detection

The change of pavement deflection is a complex process with multiple factors. The material properties, structural composition type, compaction status, compaction degree, temperature and humidity environment, climatic conditions, traffic composition, environmental conditions during testing, and the equipment used in each layer of the roadbed and pavement all have a great impact on the size of the deflection [5]. After the completion of the project, the road surface deflection was detected, and the road surface deflection was detected for two consecutive years in the fourth and fifth years of the project completion. In order to reduce the difference in the detection data, when the road surface deflection detection is carried out, the control detection conditions are kept consistent, and a measuring point is set every 15 m in the longitudinal direction. The comparison of the detection data of surface deflection is shown in Fig. 4.

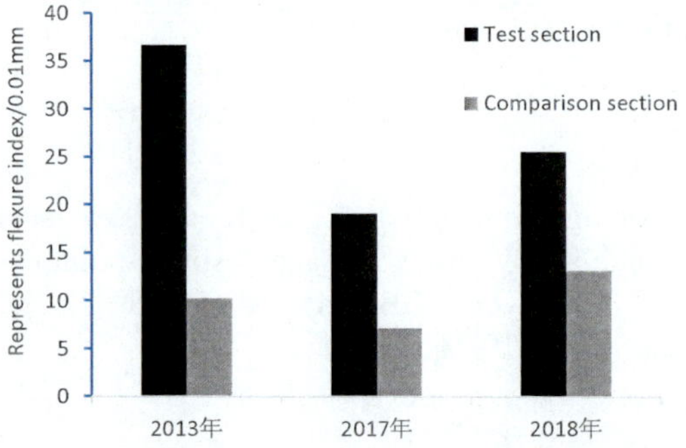

Fig. 4. Comparison chart of surface deflection value

(1) DFI of the pavement in CS is 33.7 (0.01 mm), and DFI of the pavement in TS is 43.2 (0.01 mm), and the surface deflection values (SDV) all meet the design requirements. The three inspection results show that SDV of TS is obviously greater than that of CS.
(2) In 2017, SDV of TS and CS reached the minimum, the deflection value (DV) of TS decreased by 48%, and DV of CS decreased by 62.3%. Under the effect of vehicle load, the structural layer is gradually compacted. The strength of the graded gravel layer and SRB material both increase with age, which leads to a gradual decrease in pavement deflection and eventually reaches a minimum.
(3) In 2018, DV of the two sections of the surface layer began to increase. DV of TS increased by 35.6% compared with 2017, and DV of CS increased by 83.3%. Due to the repeated rolling of the vehicle load, 4 to 5 years after the completion of the road, the growth of the strength of the base layer slows down and gradually tends

to a relatively stable state. The repetitive action of vehicle load plus the influence of factors such as strength unevenness caused by changes in water and temperature conditions will cause the microscopic defects inside the structure to expand due to local stress concentration, and gradually appear small-scale local damage. As a result, the overall stiffness of the pavement structure decreases and DV of the pavement increases.

5.3 Investigation of Pavement Cracks

In 2017 and 2018, the development of road cracks on the Street was tested and statistics, as shown in Table 8:

Table 8. Comparison of the cracks

Road section	Detection time	Number of cracks (bars)	Average number of cracks in 100 m mileage (strip)	Average number of newly added cracks in 100 m mileage (Article)	Total crack length (M)	Average total crack length of 100 m mileage (m)	Average length of newly added cracks in 100 m mileage (m)	Crack width (mm)
Comparison segment	2017.6	26	10	4.23	167.5	64.42	25.465	4–10
	2018.5	37	14.23		233.7	89.88		4–13
test section	2017.6	7	2.33	1	22.9	7.63	5	3–4
	2018.5	10	3.33		37.9	12.63		3–6

The cracks in the comparison section are obviously more serious than those in TS, which are manifested in the large number of cracks, long length and large width. The difference in the number of cracks per 100 m between CS and TS is about 5 times, and the difference in crack length is 7 to 8 times. The structural application of GBSB and LB of CSRM has effectively delayed the occurrence and development of cracks.

It can be seen that reasonable pavement structure combination is an effective way to improve the application effect of GBSB. Generally speaking, its graded aggregate is significantly lower than that of SRB materials, such as cement and other stabilized crushed stones. However, unlike semi-rigid materials, GBS has a significant nonlinearity. This characteristic makes it show a large modulus of resilience on the underlying layer with higher stiffness, so it also has sufficient resistance to stress deformation. Finally, as the upper base, GBS not only has the function of slowing down the RC of semi-rigid asphalt pavement, but also has good anti fatigue ability [2]. On the other hand, CSRM also plays a certain role as a SRB material. Compared with cement stabilized macadam, CRMA can ensure higher strength, reduce elastic modulus and foundation stiffness, and can significantly reduce cracks in urban roads without heavy traffic pressure.

6 Conclusion

Through the design of inverted structural materials and mix proportion of LB of SRB+UB of GBS, and the summary of its engineering application and detection tracking, the following conclusions are drawn:

(1) under the condition of full compaction, GBSB has good bearing capacity. When it is applied between the SRB and the asphalt surface course, it can effectively delay the generation of RC.
(2) In the structural combination design, it is necessary to ensure the high strength of the subbase, so that the nonlinearity of the GBS can be brought into full play, showing sufficient resistance to stress and ability of reducing deformation.
(3) The investigation results of pavement deflection and crack after six years show that the asphalt pavement structure of LB of semi-rigid and UB of GBS is more effective than CSG base in urban roads with less traffic pressure, which can delay the generation of RC on asphalt surface and improve the service life of pavement.

References

1. quan, Z.: Study on influence of graded macadam thickness on stress and service life of flip-top asphalt pavement structure. Jiangsu Province Highway Society, "modern traffic technology" editorial department. Collection of academic papers of Jiangsu Highway Society (2017). Jiangsu Provincial Highway Society, p. 6 (2018)
2. Zepeng, W.: Analysis on construction technology of graded macadam base of asphalt pavement. China Road **13**, 104–105 (2019)
3. Xie Xianggen, W., Shanzhou, H.K.: Research on the application of graded gravel base on expressways. China Foreign Highways **30**(02), 208–212 (2010)
4. Xuejun. D.: Roadbed and Pavement Engineering, 3rd edn., vol. 3, pp. 274–276. People's Communications Press, Beijing (2010)
5. Long, W., Yaodong, S.: Types and compaction methods of graded gravel structure materials. J. Beihua Univ. (Nat. Sci. Edn.) **5**(4), 368–372 (2004)
6. Junzhi, W., Sheng, W.: Construction quality control of flexible pavement graded gravel subbase. Guangdong Highway Commun. **21**(4), 77–79 (1995)
7. Xing, M., Shixiu, M., Binggang, W.: Determination of the reasonable range of critical sieve holes for graded crushed stone based on shear performance. J. Traffic Transport. Eng. **5**(4), 27–31 (2005)

Investigation on the Mechanical Properties of Low Plasticity Clay Contaminated with Engine Oil

Khalid Riyadh Omar(✉), Behzad Fatahi, and Lam Dinh Nguyen

School of Civil and Environmental Engineering, University of Technology Sydney (UTS), Sydney, Australia
khalid.r.omar@student.uts.edu.au

Abstract. Oil contamination presents a challenge for geotechnical engineers to remedy and improve the properties of the contaminated ground. Many researchers have shown that the behaviour of granular soil is highly affected by the level of oil contamination. However, studies on the mechanical properties of the oil-contaminated clay remain limited. Hence, this paper aims to investigate the effect of engine oil on the behaviour of clayey soil, particularly its impacts on the level of oil contamination are investigated and discussed. Extensive laboratory experiments are conducted on Kaolin clay mixed with various amounts of engine oil up to 16% and experimental results in terms of Atterberg limits, unconfined compression strength and small strain shear modulus (G_{max}) are reported. It is observed that the liquid limit and plastic limit of the oil-contaminated clay initially increases with the oil content up to an optimum value where further addition of oil results in a decrease in the liquid limit. In addition, the presence of oil within the clay structure results in a reduction of the shear strength and the shear modulus of the contaminated clay. Hence, this paper concludes that the engineering properties of the oil-contaminated clay can be adversely impacted by a considerable level of contamination.

Keywords: Oil contamination · Kaolin clay · Atterberg limits · Unconfined compression strength · Shear modulus

1 Introduction

The leakage of engine oil in the vicinity of storage tanks, industrial engines, and during transportation are accidental and can be significant, imposing an ecological threat to the environment. Engine oil contains base oil and other additives which are toxic when released to the environment and can contaminate the surrounding soil with significant adverse effects. The base oil consists mainly of petroleum-based hydrocarbons which is one of the frequently used organic chemicals in the petrochemical industry. Indeed, hydrocarbons are obtained by drilling, stored in tanks, and often transferred by ships and pipelines (Routh et al. 1969).

A. Tapase et al. (Eds.): GeoChina 2021, SUCI, pp. 21–32, 2021.
https://doi.org/10.1007/978-3-030-79644-0_3

According to Tuncan and Pamukcu (1992), when oil spillage occurs, hydrocarbons in liquid form infiltrate to the ground due to gravitational force and can spread in different directions by migrating to the capillary zone within the soil. The saturation of soil by hydrocarbons which consists of different chemical characteristics in comparison to water has been shown to have a degrading effect on mechanical, filtration, and retention characteristics of the soil (e.g. Nazir 2011). Where this hydrocarbon saturation or contamination accumulates, several undesirable incidents such as the settlement of foundations resulting in cracks within the structures or pipelines, unexpected settlements, and even unrecoverable damage to the soil environment can occur. The size of contamination is greatly dependent on the chemical composition of the infiltrated hydrocarbon and its amount as well as the initial properties of the intact soil as reported by Khamehchiyan et al. (2007).

There have been several studies in the literature to study the effects of industrial oil on the geotechnical properties of both coarse and fine grained soils (Khamehchiyan et al. 2007; Khosravi et al. 2013; Khodary et al. 2018; Safehian et al. 2018). It has been shown that various geotechnical characteristics of soils including strength, permeability, Atterberg limits, and optimum water contents are significantly impacted as compared to the uncontaminated soils. Rajabi and Sharifipour (2018) observed that the shear modulus of oil-contaminated sand increased due to the effect of weathering, particularly in the first six months after contamination. Moreover, Khodary et al. (2018) suggested that the physical interactions between the soil particles and the contaminants led to a reduction in the permeability of granular material where the pore space within the soil matrix decreased due to the migration of foreign particles in the contaminant. There are many techniques available for improvement of soft clays (e.g. Azari et al. 2016; Le et al. 2017; Nguyen et al. 2017; Parsa-Pajouh et al. 2016; Nguyen and Fatahi 2016), however, associated environmental aspects need to be investigated further. From the experimental viewpoint, a study conducted on oil-contaminated sand performed by Cook et al. (1992) has shown that the compaction properties remained intact while a decrease in friction angle was observed. A similar study conducted by Shin and Das (2001) reported a significant reduction in the bearing capacity of the oil-contaminated sand.

A study on the microscopic properties of oil contaminated clay conducted by Safehian et al. (2018) reported a decrease in the soil cohesion, internal friction angle and unconfined compressive strength, while the compression coefficient increased with increasing oil content. Most recently, Rajabi and Sharifipour (2019) performed a comprehensive review of the effect of hydrocarbon contamination on geotechnical properties of various types of soils, including Atterberg limits, maximum dry density, consolidation, and friction angle. It was concluded that while the permeability and consolidation coefficients are unanimously decreased, the effect of oil contamination on Atterberg limits and unconfined compression strength remains unclear and dependent on the quality and reliability of the conducted laboratory tests. Also, Ratnaweera and Meegoda (2006) reported an increase in the Atterberg limits with increasing oil content while contrasting test results were observed in the study conducted by Jia et al. (2010). Similarly, contradictory results were reported for the shear strength of contaminated soil determined via the unconfined compression strength (UCS) tests (Khamehchiyan et al. 2007; Khosravi et al. 2013). Therefore, the understanding of the behaviour of oil contaminated fine grained soil remains limited and the authors were not aware of a consensus in the literature about the impacts of oil contamination on the strength, stiffness and plasticity characteristics as well as the correlations between them for the contaminated clays. Hence, the main objective of this paper is to perform laboratory testing to investigate the effects of engine oil on the mechanical and plasticity properties of the contaminated clay. In this study, the impacts of oil contamination on the above-mentioned characteristics and possible correlations with Atterberg limits are discussed. Extensive laboratory experiments were conducted on Kaolin clay mixed with various oil contents ranging from 0% to 16%. Experimental results in terms of Atterberg limits, the unconfined compressive strength (UCS) values and the small strain shear modulus (G_{max}) of the oil-contaminated clay are reported and discussed.

2 Test Materials and Sample Preparation Method

In this study, artificial K10 Kaolin clay, which is commercially available in large quantities, was used to evaluate the mechanical properties of the oil-contaminated clay. The K10 Kaolin clay is classified as low plasticity clay with the Plasticity Index (*PI*) of 28.5%. Further properties of the K10 Kaolin clay measured in this study are summarised in Table 1. The contaminant used in this study is the fresh engine oil named Castrol GTX 20W-50 which is frequently used in vehicle engines. This type of engine oil is in a liquid state and insoluble in water. The chemical composition consists of up to 90% of base oil which is mainly composed of hydrocarbons. The properties of the engine oil used as a contaminant in this paper are presented in Table 2 (Castrol safety data sheet).

The preparation of the contaminated sample involved mixing dry Kaolin clay in powder form thoroughly with 10% of distilled water corresponding to the hypothetical initial condition on site. The weight of the added water was calculated based on the dry weight of the clay. The mixture was kept in an air-tight plastic bag to ensure the homogeneity and moisture equilibrium within the clay particles. After 24 h of curing, the soil was mixed with different percentage of engine oil ranging from 0% to 16% (% of oil by dry weight of soil). It should be noted that the weight of added oil was calculated based on the dry weight of the soil.

Table 1. Properties of K10 Kaolin clay adopted in this study

Soil property	K10 Kaolin clay
Liquid Limit (LL)	48.0%
Plastic Limit (PL)	19.2%
Plasticity Index (PI)	28.8%
Specific Gravity (Gs)	2.66
Linear Shrinkage (LS)	7.0%
Colour	White
Soil Classification (USCS)	CL

Table 2. Properties of engine oil adopted in this study

Grade	Castrol GTX 20W-50
Density solubility	<1000 kg/m^3 (<1 g/cm^3) at 15 °C insoluble in water
Viscosity	Kinematic: 159 mm^2/s (159 cSt) at 40 °C Kinematic: 18 to 19 mm^2/s (18 to 19 cSt) at 100 °C
Ingredient name	Base oil - highly refined Varies $\geq$90 Distillates (petroleum), hydrotreated heavy paraffinic $\leq$10

For the Atterberg limits, including Liquid Limit, Plastic Limit and Shrinkage limit tests, the specimens were kept in an air-tight plastic bag for extra 14 days to ensure the homogeneity of the mixture and uniform distribution of pore water and oil in the soil matrix. On the other hand, when the sample was prepared for the unconfined compressive strength (*UCS*) tests, the oil-clay mixture after 24 h of curing was then mixed with an additional amount of water to achieve optimum moisture content and then the mixture was cured in the air-tight plastic bags for 14 days. The oil-clay mixture was then placed in the cylindrical mould ($H = 100$ mm and $D = 50$ mm) and compacted in three equal layers to achieve maximum dry density ($\rho_d = 1.64$ t/m^3) at optimum moisture content of 19.7%. It should be noted that the adopted values for maximum dry density and optimum moisture content were determined from the standard compaction test conducted on uncontaminated Kaolin clay in accordance with the Australian Standards (AS 1289.5.1.1–2017). It should be noted that all compacted samples were prepared at the same maximum dry density of $\rho_d = 1.64$ t/m^3 for the sake of consistency among all the samples allowing assessment of oil contamination effects while keeping the soil void ratio constant.

3 Results and Discussion

3.1 Atterberg Limits of the Oil Contaminated Clay

The Atterberg limit tests including the Plastic limit (*PL*), Liquid limit (*LL*) and Shrinkage limit (*SL*) were performed on Kaolin with and without oil contamination. The oil contents were chosen at 0%, 1%, 2%, 4%, 6%, 8% and 16% as suggested by other researchers to cover a range of contamination in real practice (Khamehchiyan et al. 2007; Kermani and Ebadi 2012). In this study, the Plastic limit test was conducted in accordance with the Australian Standards AS 1289.3.2.1 (2009). The plastic limit was measured by a classical test apparatus including a glass plate, a 3 mm steel rod for thickness control of the sample and small tins to quantify the moisture content. The cone Liquid limit test was conducted in accordance with the Australian Standards AS 1289.3.9.1 (2015) where the cone penetrometer was used and the depth of penetration was recorded. The liquid limit of the sample was determined at the moisture content corresponding to a cone penetration of 20 mm at which the soil state changed from plastic to liquid. As suggested by Herrick and Jones (2002), the dependency of the cone Liquid limit test results on the test operator is expected to be negligible which is the advantage of this method in comparison to Casagrande device. The results for Liquid limit (*LL*) and Plastic limit (*PL*) of the oil contaminated Kaolin clay are reported in Fig. 1. Referring to Fig. 1 it is evident that increasing oil content initially resulted in an increase in the *LL* of the oil-contaminated soil while a further increase beyond an optimum amount of oil led to a reduction in *LL*. Similar observations were also made for the plastic limit of the adopted oil contaminated soil. The results of increasing *LL* and *PL* conducted in this study are consistent with those results reported by Kermani and Ebadi (2012).

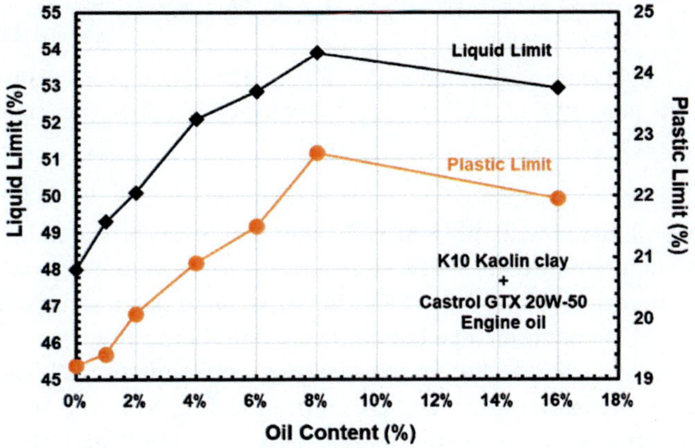

Fig. 1. Effect of oil contents on the Liquid Limit (LL) and Plastic Limit (PL) of the contaminated Kaolin clay

In addition, when the amount of oil increased from 0% to 8%, there was a 12.3% increase in *LL* and 18.0% increase in *PL* while a reduction of 1.8% in *LL* and 3.4% in *PL* were observed when the oil content further increased to 16%. Hence, the effect of oil contamination is more significant on the plastic limit as compared to the liquid limit of the Kaolin clay.

An important soil characteristic parameter arising from the *LL* and *PL* of a particular soil is the Plasticity Index (*PI*) which expresses the range of moisture content where the soil exhibits plastic behaviour and it is determined as $PI = LL - PL$. Figure 2 showed an increase in *PI* when more engine oil was added to the soil up to 6% oil content, followed by a slight reduction in *PI* with a further increase in the amount of oil. It was observed that the plasticity of the soil with 0% oil content increased up to 8.9% as compared to the highest *PI* value of 31.35% corresponding to the 6% oil contaminated sample while the highest values of *LL* and *PL* occurred at 8% oil content.

In addition, results of the Linear Shrinkage (*LS)* tests conducted on the oil contaminated clay are reported in Fig. 3. The *LS* of the soil pertains to the percentage of shortening of the dried soil bar with respect to its initial length at the moisture content equal to *LL*. The results of *LS* tests are mainly used to understand the potential of the soil in experiencing volume change due to moisture content variations (e.g. shrinkage and swelling). The *LS* tests in this study were performed in accordance with the Australian Standard AS 1289.3.4.1 (2008) using the linear shrinkage test apparatus, which included a 250 mm long and 25 mm wide half-cylinder metal mould. The *LS* moulds were filled

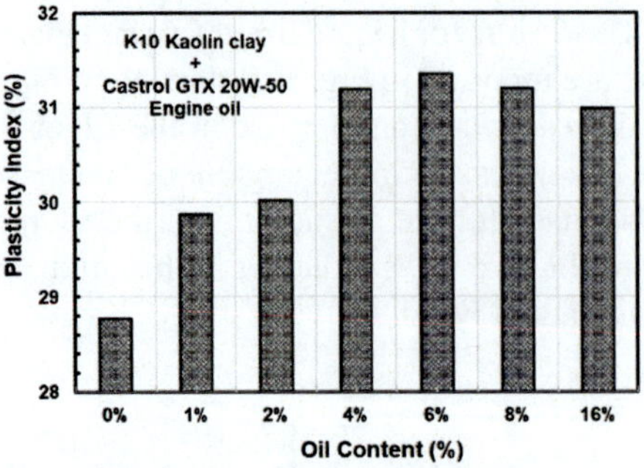

Fig. 2. Effect of oil content on the Plasticity Index (PI) of contaminated Kaolin clay

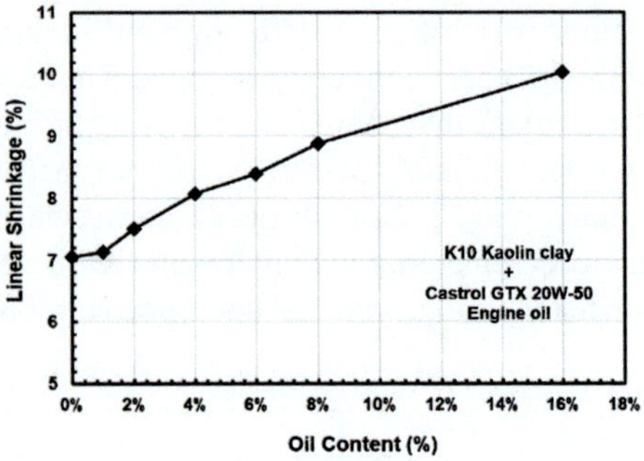

Fig. 3. Effect of oil content on the Linear Shrinkage (LS) of the contaminated Kaolin clay

with the Kaolin soil mixed with different oil contents at their corresponding liquid limits and kept at room temperature for 24 h and then oven-dried. Figure 3 shows the variation of *LS* with oil contents. Evidently, the *LS* of the oil contaminated sample increased with the increasing level of contaminants. In contrast to Atterberg Limits (i.e. *LL* and *PL*), no optimum oil content corresponding to the maximum or minimum *LS* was observed for the adopted soil and engine oil mixtures. Referring to Fig. 3, When 16% of engine oil was added to the uncontaminated soil, the *LS* increased significantly from 7% to 10% (i.e. 42.4% increase).

3.2 Unconfined Compressive Strength (*UCS*) of the Oil Contaminated Clay

The Unconfined Compressive Strength (*UCS*) of soil shows the maximum compressive stress that can be applied to an unconfined specimen and is often used to correlate to the shear strength of the material. Due to its economical and practical advantages, the *UCS* test is widely performed to measure soil strength. Moreover, the Young's (*E*) can also be estimated using the stress-strain curve result obtained from the *UCS* test. In this study, a series of *UCS* tests were conducted in accordance with the Australian Standard AS 5101.4 (2008) on the contaminated sample with oil contents of 0%, 1%, 2%, 4% and 6% while keeping the dry density constant equal to 1.64 (g/cm^3). During the sample preparation, it was observed that the soil sample with oil contents higher than 6% could not achieve the required dry density due to excessive oil contents occupying the soil voids comparable to wet soils as reported by Khamehchiyan et al. (2007). Furthermore, according to Al-Sanad et al. (1995), the addition of excessive oil contamination could lead to the compaction curve losing some of its characteristics and shape.

The stress and strain responses of the contaminated compacted clays with various oil contents are presented in Fig. 4. The stresses mobilised in the Kaolin clay increased to the peak (q_{max}) and then gradually decreased with increasing axial strain (ε) indicating the strain softening. The $q_{max} = 120.9$ kPa of the uncontaminated clay was recorded at $\varepsilon = 3.49\%$ while there was a significant loss in strength (i.e. $q_{max} = 30.2$ kPa) corresponding to $\varepsilon = 4.95\%$ when 6% oil was added to the compacted soil. It can be observed that the strain corresponding to the peak strength generally increased due to the addition of the oil, which is consistent with the experimental results reported by Safehian et al. (2018).

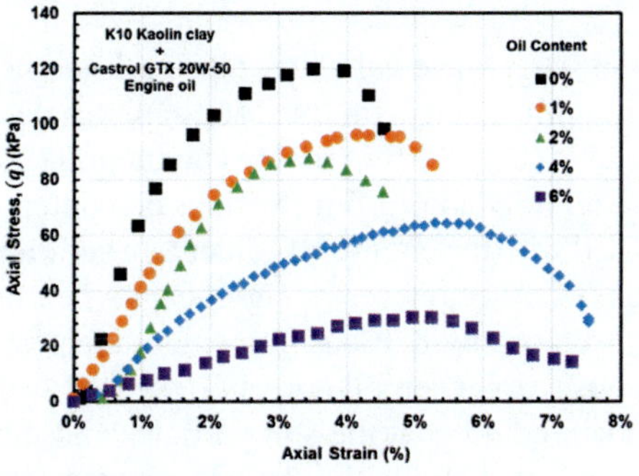

Fig. 4. Effect of oil content on the stress-strain response of contaminated Kaolin clay

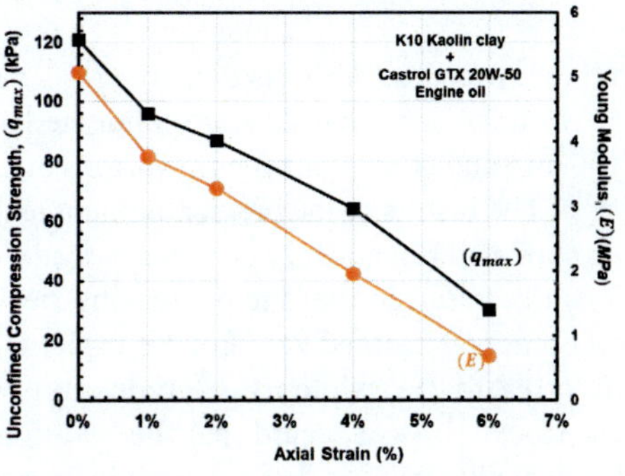

Fig. 5. Effect of oil content on the UCS and Young's modulus (E) of the contaminated Kaolin clay

Furthermore, Fig. 5 showed the variation of the *UCS* of the compacted Kaolin clay contaminated with oil. It was observed that the strength of the contaminated Kaolin clay reduced drastically by approximately 75% when 6% engine oil was added. Furthermore, from the stress and strain curve in Fig. 4 Young's modulus (E) was determined and are reported in Fig. 5. The Young's modulus was determined using the linear regression over the initial linear section of the stress-and-strain curve where the sample exhibited elastic behaviour. As shown in Fig. 5, there was a significant drop (approx. 86.4%) in the stiffness of the oil contaminated clay samples due to the addition of 6% engine oil ($E =$ 5.07 MPa for uncontaminated soil as compared to $E = 0.69$ MPa for soil contaminated with 6% oil). Hence, the strength and stiffness of the compacted clay were adversely impacted by oil contamination.

3.3 Small-Strain Shear Modulus (G_{max}) of the Oil Contaminated Clay

The small-strain shear modulus is one of the most significant and frequently used soil characteristics in geotechnical engineering where the evaluation of ground response subjected to small strains or transition from small strain to large strain is of interest such as evaluating the excavation stability and performance (Lee and Santamarina 2005). Among various methods to determine G_{max}, bender element method has shown to be highly capable, fast, and accurate (Lee and Santamarina 2005). Bender elements are made from piezoelectric ceramic and embedded in the base pedestal and the top cap which were employed in this study to estimate the shear wave velocity of the soil specimen. Bender element test was performed by applying an excitation from one element attached to one end of the sample causing a shear wave to travel within the soil sample, while the other element attached to the other end of the sample received the transmitted signal. Data were collected using a data acquisition unit, and the shear wave velocity was calculated based on the measured travel time and the distance between the two elements. Equation (1) was used to calculate the small-strain shear modulus (G_{max}) of the soil considering the measured shear wave velocity and soil density.

$$G_{max} = \rho V_s^2 \quad (1)$$

Where, ρ is the soil mass density and V_s^2 is the shear wave velocity of the soil sample.

Bender element tests were conducted on the compacted cylindrical soil samples (H = 100 mm and D = 50 mm) with various oil contents (i.e. 0%–6%) before conducting *UCS* tests. The results of the bender element test are reported in Fig. 6 for various oil contents. As shown in Fig. 6, G_{max} progressively decreased with the engine oil content. In addition, it can be inferred that the decreasing trend of G_{max} became more significant when the oil content exceeded 2%. For example, G_{max} reduced by 8.3% as oil content increased from 0% to 2%, while 61.2% reduction was observed when oil content increased from 2% to 6%. It is expected that the addition of oil, could coat the soil particles and thus lubricate the frictional contact between particles and allowing the slip between particles easily and thus resulting in lower G_{max}.

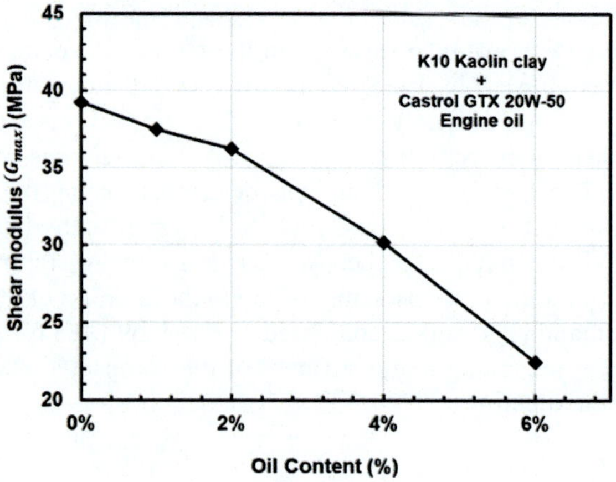

Fig. 6. Effect of oil content on the shear modulus (G_{max}) of contaminated Kaolin clay

4 Discussion

It was observed from Fig. 1 that the *LL* and *PL* of the contaminated samples increased to an optimum value (i.e. 8% oil content) which was then followed by a decline when a further amount of oil was added to the clay. As explained by Kermani and Ebadi (2012), the plasticity behaviour of the oil contaminated soil is mainly controlled by the double-layer water where the water molecules are attracted by the hydrogen atoms of the clay due to the dipolar characteristic of the water (positive and negative charges of water molecules) to form a hydrogen bonding with the clay particles. Since the oil encompasses the clay particle, an additional amount of water is needed for the clay to exhibit its plastic properties, thus *PL* rises with the increase of the oil content Ota

(2013). In addition, the remaining water outside the double-layer water (i.e. the pore space) is called the free water which determines the liquid limit of the soil. Accordingly, the amount of water required by the soil to approach its liquid state increases with the oil content. The increases in *PL* and *LL* result in the increase in Plasticity Index (*PI*) of the soils as can be shown in Fig. 2. However, both *LL* and *PL* reduce their values as oil content exceeds 8%. This behaviour can be described based on the viscous nature of the contaminating oil as elaborated by Khosravi et al. (2013). When soils are heavily contaminated with oil, a thick layer of contaminant serves as viscous interface layer that preserves the clay structure and helps the particles to slip over each other. Therefore, the amount of water needed for the contaminated clay to reach its *LL* and *PL* decreases. Moreover, due to the increase in *PI*, the behaviour of Kaolin clay has transitioned from a low plasticity clay (*CL*) to a high plasticity clay (*CH*) as clearly shown in the plasticity chart (Fig. 7). Hence, in this study, there was a critical oil content range (i.e. 4%–8%) where Kaolin clay changed its plasticity behaviour due to the significant effect of oil contamination.

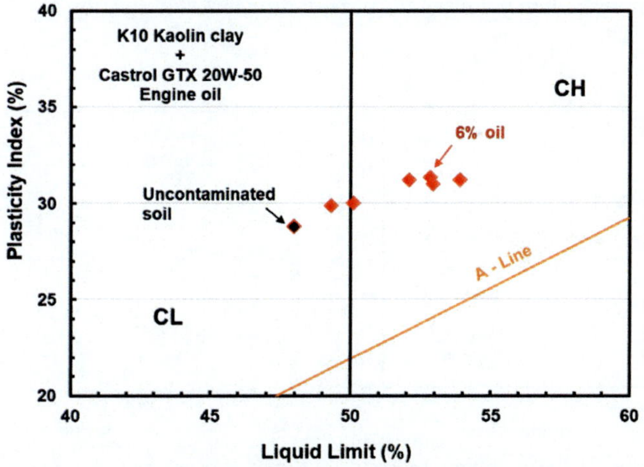

Fig. 7. Effect of oil content on the classification of contaminated Kaolin clay

The effects of the oil contamination on the *UCS* and small-strain shear modulus of the contaminated soil were summarised in Figs. 4, 5, and 6. It was observed that the increase in oil content has a diminishing effect on soil strength. This is mainly attributed to the fact that the addition of oil contaminants coated the clay particles, which reduced the inter-particle forces and thus the particles could slip more easily under the applied external load as also elaborated by Khamehchiyan et al. (2007). In addition to the effect of the mechanical interaction, the hydrocarbon compounds acting as a nonpolar fluid might increase the intermolecular bond between clay particles, while this additional bond between the particles cannot overcome the tendency of the particle slippage where viscous layers are surrounding the particles (Onyelowe 2015). Moreover, as oil content increased, the void saturation increased which led to lower soil suction and consequently, lower shear strengths were observed. Figure 8 compares the correlation of the plasticity index obtained from the Atterberg limit tests and the shear strength and the small-strain shear modulus measured by exploiting the shear strength ratio (i.e. $q_r = q_{max}/PI$) and the small-strain shear modulus ratio (i.e. $G_r = G_{max}/PI$). As shown in Fig. 8, both q_r and G_r decreased with the oil content. However, the effect of oil contamination was more prevalent on q_r (i.e. a reduction of 77%) as compared to a reduction of 47% in G_r. It can be observed that the shear strength and the small-strain shear modulus are inversely proportional to the plasticity of the contaminated clay. Figure 9 illustrated the correlation of the normalised shear strength ($q_n = q_{max}/E$) and the normalised shear modulus ($G_n = G_{max}/E$) with increasing oil content. The results show that q_n ranged between 23.85 and 44.06, while G_n changed between 7.74 and 32.70, for the considered oil contamination levels. These ranges can be used by practicing engineers to estimate Young's modulus and unconfined compressive strength of compacted oil contaminated clays based on the shear wave velocity or small-strain shear modulus obtained from non-destructive tests. It should be noted that it is expected that the migration of engine oil into the soil skeleton can saturate the voids and thus increases the soil plasticity, resulting in a loss of strength and stiffness of the clay. This presents a significant challenge for practicing engineers when planning the remediation and stabilisation of oil contaminated clays.

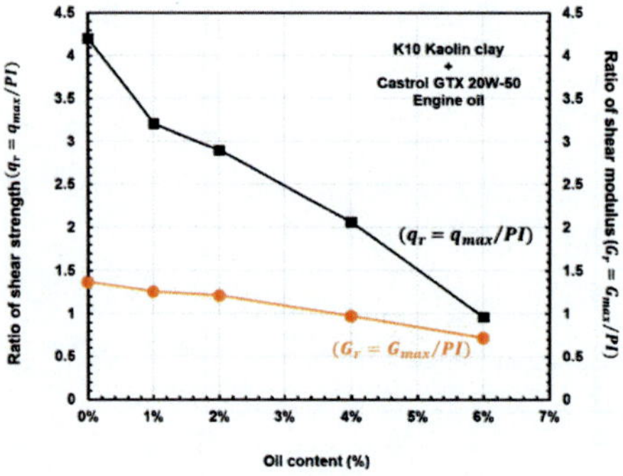

Fig. 8. Variations of shear strength (q_r) and shear modulus (G_r) ratios with oil content

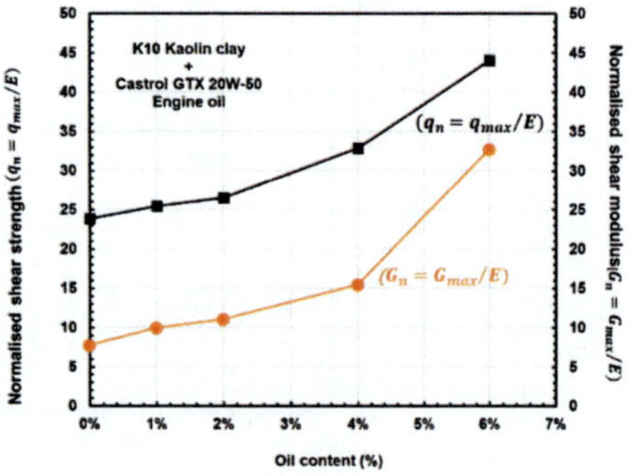

Fig. 9. Variations of normalised shear strength (q_n) and normalised shear modulus (G_n) with oil content

5 Conclusions

The effect of oil contamination on clay has imposed significant challenges for geotechnical engineers. The contamination of sandy soil has been studied extensively while investigation on the engineering behaviour of the oil-contaminated clay remains limited. In this paper, extensive laboratory tests were conducted on Kaolin clay mixed with various percentages of engine oil, ranging from 0% to 16%, to investigate the effect of oil content on the plasticity characteristics and shear strength of the clay. The experimental program included the Atterberg limit tests, unconfined compressive strength (UCS) test and the bender element test. It was observed that the addition of engine oil increased the Atterberg limits and the plasticity index of clay. Indeed, the oil contaminated Kaolin clay transitioned its classification from low plasticity to high plasticity clay when oil was added beyond an oil content threshold. Moreover, the increase in plasticity index could be correlated with the significant reduction in the shear strength and stiffness of the contaminated soil. The results evidently show that the shear strength and stiffness of the contaminated soil were adversely impacted due to the oil coating the surface of the clay particles and basically acting like a lubricant. However, the mechanism of oil and clay interaction at microscopic scale requires further investigation.

References

Al-Sanad, H.A., Eid, W.K., Ismael, N.F.: Geotechnical properties of oil-contaminated Kuwaiti sand. J. Geotech. Eng. **121**(5), 407–412 (1995)

Azari, B., Fatahi, B., Khabbaz, H.: Assessment of the elastic-viscoplastic behavior of soft soils improved with vertical drains capturing reduced shear strength of a disturbed zone. Int. J. Geomech. **16**(1), B4014001 (2016)

Cook, E.E., Puri, V.K., Shin, E.C.: Geotechnical characteristics of crude oil-contaminated sands. In: Proceeding of 2nd International Offshore Polar Engineering Conference San Francisco, USA, pp. 384–387 (1992)

Head, K.H., Epps, R.J.: Manual of Soil Laboratory Testing, (vol. 1, No. 2). Pentech Press, London (1980)

Herrick, J.E., Jones, T.L.: A dynamic cone penetrometer for measuring soil penetration resistance. Soil Sci. Soc. Am. J. **66**(4), 1320–1324 (2002)

Jia, Y.G., Wu, Q., Meng, X.M., Yang, X.J., Yang, Z.N., Zhang, G.C.: Case study on influences of oil contamination on geotechnical properties of coastal sediments in the Yellow River Delta. In: Chen, Y., Zhan, L., Tang, X. (eds.) Advances in Environmental Geotechnics, pp. 767–771. Springer, Heidelberg (2010). https://doi.org/10.1007/978-3-642-04460-1_94

Kermani, M., Ebadi, T.: The effect of oil contamination on the geotechnical properties of fine-grained soils. Soil Sediment Contamin. Int. J. **21**(5), 655–671 (2012)

Khamehchiyan, M., Charkhabi, A.H., Tajik, M.: Effects of crude oil contamination on geotechnical properties of clayey and sandy soils. Eng. Geol. **89**(3–4), 220–229 (2007)

Khodary, S.M., Negm, A.M., Tawfik, A.: Geotechnical properties of the soils contaminated with oils, landfill leachate, and fertilizers. Arab. J. Geosci. **11**(2), 1–17 (2018). https://doi.org/10.1007/s12517-017-3372-7

Khosravi, E., Ghasemzadeh, H., Sabour, M.R., Yazdani, H.: Geotechnical properties of gas oil-contaminated Kaolinite. Eng. Geol. **166**, 11–16 (2013)

Le, T.M., Fatahi, B., Khabbaz, H., Sun, W.: Numerical optimization applying trust-region reflective least squares algorithm with constraints to optimize the non-linear creep parameters of soft soil. Appl. Math. Model. **41**, 236–256 (2017)

Lee, J.S., Santamarina, J.C.: Bender elements: performance and signal interpretation. J. Geotech. Geoenviron. Eng. **131**(9), 1063–1070 (2005)

Nazir, A.K.: Effect of motor oil contamination on geotechnical properties of over consolidated clay. Alex. Eng. J. **50**(4), 331–335 (2011)

Nguyen, L., Fatahi, B.: Behaviour of clay treated with cement & fibre while capturing cementation degradation and fibre failure–C3F model. Int. J. Plast. **81**, 168–195 (2016)

Nguyen, L., Fatahi, B., Khabbaz, H.: Development of a constitutive model to predict the behavior of cement-treated clay during cementation degradation: C3 model. Int. J. Geomech. **17**(7), 04017010 (2017)

Onyelowe, K.C.: Pure crude oil contamination on amaoba lateritic soil. Electron. J. Geotech. Eng. **20**(3), 1129–1142 (2015)

Ota, J.O.: The Effect of light crude oil contamination on the geotechnical properties of kaolinite clay soil. Doctoral dissertation, Anglia Ruskin University (2013)

Parsa-Pajouh, A., Fatahi, B., Khabbaz, H.: Experimental and numerical investigations to evaluate two-dimensional modeling of vertical drain–assisted preloading. Int. J. Geomech. **16**(1), B4015003 (2016)

Rajabi, H., Sharifipour, M.: Influence of weathering process on small-strain shear modulus (Gmax) of hydrocarbon-contaminated sand. Soil Dyn. Earthq. Eng. **107**, 129–140 (2018)

Rajabi, H., Sharifipour, M.: Geotechnical properties of hydrocarbon-contaminated soils: a comprehensive review. Bull. Eng. Geol. Environ. **78**, 3685–3717 (2019)

Ratnaweera, P., Meegoda, J.N.: Shear strength and stress-strain behavior of contaminated soils. Geotech. Test. J. **29**(2), 133–140 (2006)

Routh, J.I., Eyman, D.P., Burton, D.J.: Essentials of General, Organic and Biochemistry, p. 130, W. B. Sunders Company, Philadelphia (1969)

Safehian, H., Rajabi, A.M., Ghasemzadeh, H.: Effect of diesel-contamination on geotechnical properties of illite soil. Eng. Geol. **241**, 55–63 (2018)

Shin, E.C., Das, B.M.: Bearing capacity of unsaturated oil-contaminated sand. Int. J. Offshore Polar Eng. **11**(03), 220–227 (2001)

Tuncan, A., Pamukcu, S.: Predicted mechanism of crude oil and marine clay interactions. In: Proceedings of Environmental Geotechnology, Taylor and Francis, Turkey, pp. 205–212 (1992)

Preliminary Observations of Astronomical Coordinates by the SDUST/NAO Digital Zenith Tube

Jiajia Yuan[1], Jinyun Guo[1,2(✉)], Yi Shen[1], Jie Dai[1], Xin Liu[1], and Qiaoli Kong[1,2]

[1] College of Geodesy and Geomatics, Shandong University of Science and Technology, Qingdao 266590, PR China

[2] State Key Laboratory of Mining Disaster Prevention and Control Co-Founded by Shandong Province, Ministry of Science and Technology, Shandong University of Science and Technology, Qingdao 266590, PR China

Abstract. The digital zenith tube (DZT) is developed based on the astrogeodetic and photogrammetric theory. One prototype of DZT integrating techniques of GPS and CCD photogrammetry has been developed jointly by Shandong University of Science and Technology (SDUST) and National Astronomical Observatories (NAO) of Chinese Science Academy, China. It can automatically estimate astronomical coordinates in real time by CCD star image processing and GPS data processing. The DZT's errors are analyzed in three aspects, namely, the instrument error, observation error and data processing error. Reliability and accuracy of DZT integrating GPS and CCD are verified by the practical field operations. Observing results on many unknown points by the repeated observations of each point show the internal precision of the astronomical latitude and longitude. Observing results on several known points and the comparison with the EGM2008-derived results verify the external accuracy of the DZT.

Keywords: Digital zenith tube · Astronomical coordinate · GPS · CCD photogrammetry · Astrogeodetic method

1 Introduction

The astronomical meridian plane is formed with one plumb line and the Earth rotation axis in the astronomical coordinate system. The angle between the plane and the Greenwich meridian plane is known as the astronomical longitude, and the angle between the plumb line and the equatorial plane is called as the astronomical latitude. It is of great significance for the study of astronomy, geodesy, geophysics and other geosciences to precisely measure astronomical coordinates (Kühtreiber 1999; Gerstbach and Pichler 2003; Kong et al. 2010; Abdujabbarov et al. 2016).

In the past, astronomical theodolites like T4 and J05 were usually used to precisely estimate the astronomical coordinate through the traditional astrogeodetic method. It takes a long time and a lot of resources to implement the field astronomical observation using this old theodolite because of its complex manual operation, and the final results

A. Tapase et al. (Eds.): GeoChina 2021, SUCI, pp. 33–46, 2021.
https://doi.org/10.1007/978-3-030-79644-0_4

are of low precision under effects of the environment, instrument and personal errors (Hirt 2006; Gaivoronskii et al. 2015). By the 1990s, the mobile photographic zenith camera system, mainly used for mobile measuring of plumb line, has been applied in some countries of Europe and America. These instruments also adopt the analog photographic technique. Therefore, it will take 3 to 5 h to process the analog photos by the experienced engineers through the manual or semi-automatic image coordinate treatment with the help of the coordinate measurer (Hirt et al. 2010; Hirt 2010). The CCD-based total stations such as Q-Daedalus (Hauk et al. 2017) also can be used to measure the astrogeodetic deflection of the vertical (DOV) when a digital zenith camera is not available or deployable (e.g., when observations are needed in rugged terrain away from roads). Although, DOVs from Q-Daedalus are of high quality, but not quite as accurate as those from digital zenith cameras (Hauk et al. 2017). It is still an urgent problem to find a new method to automatically measure accurate astronomical coordinates.

The digital zenith tube (DZT) for measuring astronomical coordinates is developed based on the analog zenith tube with the rapid development of Global Navigation Satellite Systems (GNSS), photogrammetry, charge-coupled device (CCD), digital image processing and computer techniques (Hirt and Seeber 2008; Hanada et al. 2012; Tian et al. 2014). The new method can make up for the deficiencies of the traditional astrogeodetic methods with the improvement of star image processing, stellar area auto-searching, determination of energy center, star catalogue processing, coordinate transformation, and star identification and matching (Abele et al. 2012; Song 2012; Petrova and Hanada 2013; Wang et al. 2014; Dai et al. 2015).

The DZT is first studied by Hannover University in Germany since the 1980s, and then the University of Munich and Vienna began to study the digital zenith technology, but the general accuracy is not high, in $1''$–$2''$ (Song 2012). Around 2004, Hannover University in Germany developed a digital zenith on the basis of the original analog zenith and achieved an accuracy of $0.1''$–$0.3''$ (Hirt and Seeber 2008). Then, it cooperated with the Swiss University of Technology Zurich to develop another set of more automated digital zeniths system (Halicioglu et al. 2012). The research on the DZT started relatively late in China, and the best accuracy is about $0.3''$–$0.5''$ (Song et al. 2006; Song 2012; Wang 2014; Wang et al. 2014; Dai et al. 2015).

One prototype of DZT has been developed jointly by the Shandong University of Science and Technology (SDUST) and the National Astronomical Observatories (NAO) of Chinese Science Academy, China (Tian et al. 2014). Like other DZT (Hirt and Seeber 2008; Hirt et al. 2010; Zariņš et al. 2016), our design is mainly composed of optical telescope, CCD, GPS, leveling system, control system and data processing system. It can perform astronomical and geodetic survey observations with high precision, high automation and real-time, and has very important application value in geodetic surveying.

In the paper, the astronomical coordinate errors will be analyzed based on the principle and hardware of the digital zenith tube. The field observations are made to test and verify the reliability and accuracy of the astronomical coordinates measured by the digital zenith tube.

2 Basic Principles

The digital zenith tube with its optical axis pointing to the zenith is set up on one station to make star photos in the zenith sky field, and its details can be found in Hirt et al. (2010). The star image coordinates can be computed in the CCD image coordinate system. One GPS receiver can be used to precisely measure the geodetic coordinate and determine the CCD exposure epoch. The information of stars in the zenith sky field can be obtained from the star catalogue by identification and matching of stars in the zenith tangent plane and the CCD image. Then the zenith equatorial coordinate (δ, α) can be estimated with the iterative least squares method and transformed to the astronomical coordinate (Φ, Λ) which is combined with the geodetic coordinate (φ, λ) to get the DOV on the measuring point (Hirt and Flury 2008; Guo et al. 2011a).

The astronomical coordinate can be converted from the zenith equatorial coordinate by

$$\begin{aligned} \Phi &= \delta \\ \Lambda &= \alpha - GAST \end{aligned} \tag{1}$$

Where, *GAST* is the Greenwich apparent sidereal time. The Helmert DOV (Hirt 2006; Vittuari et al. 2016) is

$$\begin{aligned} \xi_H &= \Phi - \varphi \\ \eta_H &= (\Lambda - \lambda)\cos\varphi \end{aligned} \tag{2}$$

The DZT prototype integrating GPS and CCD photogrammetry techniques in China has been developed jointly by SDUST and NAO (Tian et al. 2014), seeing Fig. 1. It is mainly composed of CCD imaging sensor, precise electronic level meter, optical telescope, GPS receiver, servo motor, automatic control system, data processing software and other components. Details about this instrument are detailed in Tian et al. (2014).

3 Error Analysis of Astronomical Coordinate

The digital zenith tube inevitably has some errors in mechanical machining, parts assembly, electronic control and observation and data processing, which may affect the final result of astronomical coordinate. It is necessary to analyze and correct these errors.

3.1 Instrumental Errors

The joint surface between the rotating part and the fixed part of the mechanical turntable is not a real plane. The turntable error can cause the spatial orientation change of the rotating platform axis during the rotation of the mechanical turntable. The roughness of the joint surface is 1 μm tested by one dial indicator. Experimental results show that the turntable error can be eliminated by the two observations with 0° and 180° of the DZT telescope and the tilt correction by the high-precision two-dimensional electronic level meter.

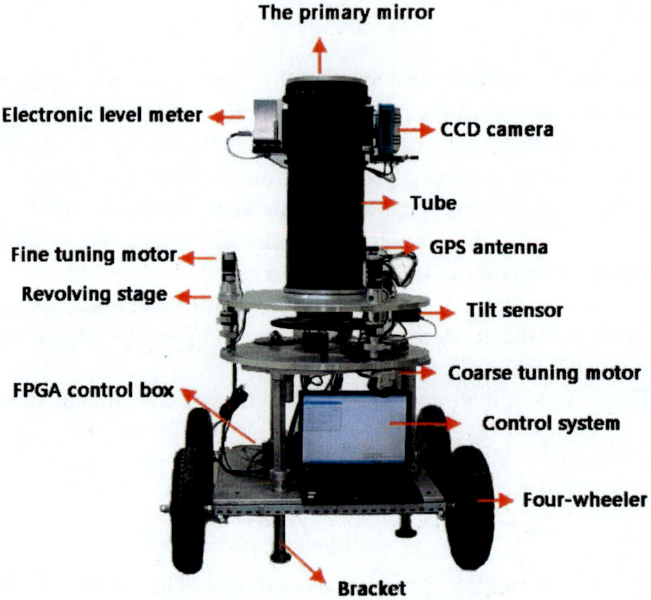

Fig. 1. Prototype of digital zenith camera integrated GPS and CCD in China

It is not possible to achieve perfect alignment between the mechanical rotation axis and the optical axis in the instrument assembly, which can bring the misalignment error. The direction of the rotation axis can be obtained by averaging the observed data of 0° and 180° positions. The misalignment error can also be eliminated by the two observations of 0° and 180° the digital zenith camera telescope (Song 2012; Wang 2014).

3.2 Observing Errors

The spatial direction of the digital zenith tube will change under influence of the external condition, such as temperature, atmospheric pressure, wind and unstable foundation during the observing procedure. If the effect of external environment on the spatial direction of the instrument axis can be ignored between two exposures in one observation, the final result will not be affected by the observing error. Otherwise, the observing error can be eliminated by two observations of 0° and 180° while the instrument tilt status changes for these two exposures in one observation. These errors are like the turntable error.

The electronic level meter used in the digital zenith tube which has the zero error and the temperature drift cannot measure the absolute tilt of the instrument, but can show the relative tilt for two exposures in one observation. In general, there is one-time interval of 20 s between these two exposures. The temperature drift of the level meter caused by the temperature change in the small interval still exists. In addition, the recording epoch by the level meter is not strictly synchronous with the CCD exposure epoch, which causes the small difference between the recorded tilt and the real instrumental tilt at the CCD exposure epoch (Guo et al. 2011b). So, the observing error is caused by the non-synchronization.

The test results of the electronic level meter show that the observing error caused by the temperature drift between the two exposures of one observation is less than 0.005″ which can be neglected. The tilt meter measurements change in the observing interval of 20 s, which is related to the surrounding environment. The effect of strong environment interference on observation cannot be ignored. The observation site should be located on the area with less environmental interference.

The observation at directions of 0° and 180° as one characteristics of the digital zenith tube is beneficial to eliminate multi-error effects. It is very difficult to precisely determine the observing place, which can cause the positioning error of the rotational axis. The rotational axis positioning accuracy is better than 2 "which can cause the astronomical observation error of less than 0.0001".

3.3 Data Processing Error

The negative model of observed CCD image represents a transforming relation between the reference star coordinate and the image coordinate. The negative model error directly affects the final measuring result.

The linear model with eight parameters or twelve parameters is generally adopted for the CCD image captured by the digital zenith tube with the view field of less than 2°. The test shows that the accuracy of the eight-parameter model is higher than that of the twelve-parameter model, which indicates that the eight-parameter model is more suitable for the CCD image processing of the digital zenith tube.

Polar motion is the motion of the Earth's rotation axis relative to the crust (Song et al. 2006; Cambiotti et al. 2016). Because the measurement of astronomical latitude and longitude is referenced to the instantaneous pole, the polar motion will have impact on the astronomical coordinates of ground point. The corrections of polar motion to astronomical coordinates (Φ_0, Λ_0) determined by the digital zenith tube are.

$$\begin{aligned} \Delta\Phi_p &= -\left(x_p \cos\Lambda_0 - y_p \sin\Lambda_0\right) \\ \Delta\Lambda_p &= -\left(x_p \sin\Lambda_0 + y_p \cos\Lambda_0\right) \tan\Phi_0 \end{aligned} \quad (3)$$

where, $\left(\Delta\Phi_p, \Delta\Lambda_p\right)$ are the corrections of polar motion and $\left(x_p, y_p\right)$ are the coordinates of the instant pole offered by the International Earth Rotation and Reference System Service (IERS) (Bizouard and Gambis 2011).

4 Practical Observing Results and Discussion

In order to verify the feasibility and accuracy of astronomical coordinate measured by the digital zenith tube, a lot of experiments and extensive data analysis are required. The internal precision of astronomical coordinates can be obtained by processing and analyzing the long-time observations of the same point. The external precision can be obtained by processing and analyzing the observational data of the different points.

4.1 Verification of Internal Precision

The internal precision reflects the difference between multiple measurements on one point, which indicates the stability of the measurement. In this paper, the internal precision of the digital zenith tube is proved by practical observations on Qingdao, Wuhan, Zhengzhou, Shenzhen, etc.

4.1.1 Practical Case on Qingdao

The data acquisition and processing by the digital zenith tube were carried out in 34 nights under the good weather and suitable environment from June 6 to December 19, 2013 in SDUST at Qingdao. 275 groups of observations were captured in the field practical experiment. In each group, the average observing time interval, the average number of observations and the average number of stars were 15 min, 26 and 724, respectively. The standard deviations (STD) were 0.246 "for the astronomical latitude and 0.269" for the astronomical longitude within a group. The distribution of the STDs for these 275 groups is shown in Fig. 2.

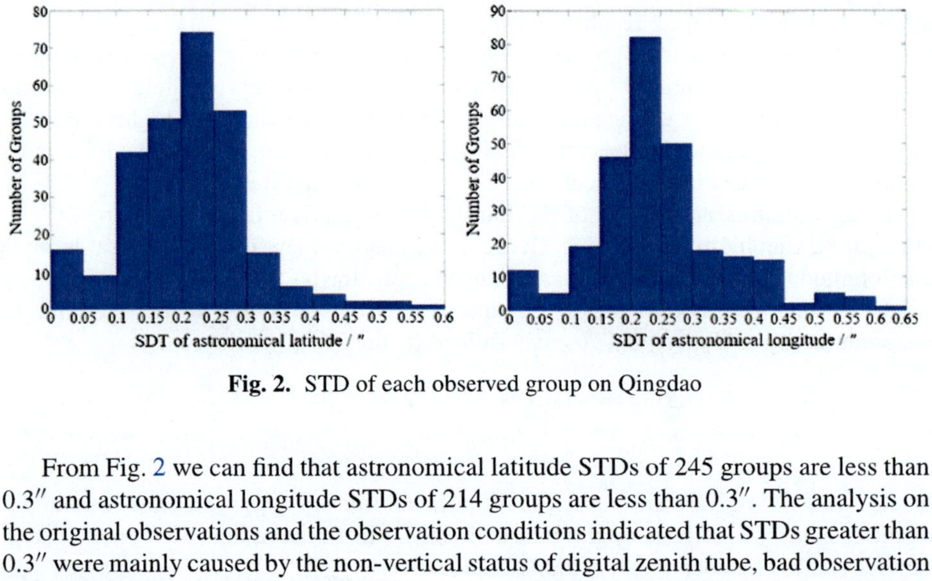

Fig. 2. STD of each observed group on Qingdao

From Fig. 2 we can find that astronomical latitude STDs of 245 groups are less than $0.3''$ and astronomical longitude STDs of 214 groups are less than $0.3''$. The analysis on the original observations and the observation conditions indicated that STDs greater than $0.3''$ were mainly caused by the non-vertical status of digital zenith tube, bad observation environment, heavy cloud or thick cloud in the zenith direction while observing. We also compute the root mean square errors of astronomical latitude and longitude in one night, which are $0.112''$ and $0.124''$, respectively.

4.1.2 Practical Case on Wuhan

The practical experiment was made on Wuhan in July 21 to 22, 2014. We construct one polygon with the fixed distance and direction to inspect the internal precision of the digital zenith tube. We selected one-point A and then determine points B, C, E and D

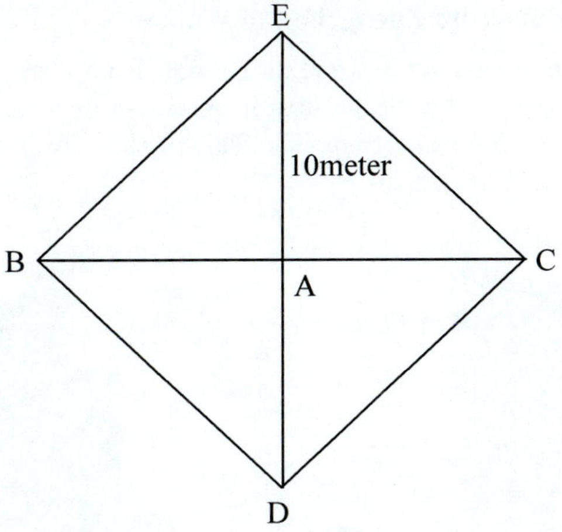

Fig. 3. Observing network on Wuhan

along the west, east, north and south directions of A with the distance of 10 m, seeing Fig. 3.

The observations were carried out on these five sites in the alphabetical turn. 2 groups of observations (15 min for each group observation) were obtained on each site. The arithmetic mean was calculated as the final astronomical coordinates of the site. Comparing the observed astronomical coordinate difference with its corresponding theoretical value, we can get the comparing results listed in Table 1. From Table 1, we can calculate the root mean squares errors (RMS) for the observed astronomical latitude and longitude are 0.053″ and 0.067″, respectively.

Table 1. Comparison of measured values and theoretical values on Wuhan

	Difference of astronomical latitude (″)		Difference of astronomical longitude (″)	
	theoretical value	Measured value	Theoretical value	Measured value
A-B	0.000	0.018	0.324	0.402
A-C	0.000	0.198	−0.324	−0.339
A-D	0.324	0.425	0.000	−0.035
A-E	−0.324	−0.327	0.000	0.186
B-C	0.000	−0.050	−0.648	−0.534
D-E	−0.648	−0.652	0.000	0.023
B-D	0.324	0.277	−0.324	−0.437
B-E	−0.324	−0.375	−0.324	−0.216
C-D	0.324	0.327	0.324	0.304
C-E	−0.324	−0.325	0.324	0.225

4.1.3 Observed Results from Zhengzhou to Wuhan

The observations were carried out on 19 feature points from Zhengzhou to Wuhan from July 15 to August 6, 2014. Total 58 groups of observations were obtained. The mean observing time is 15 min for each group. The STD of each observed group is shown in Fig. 4.

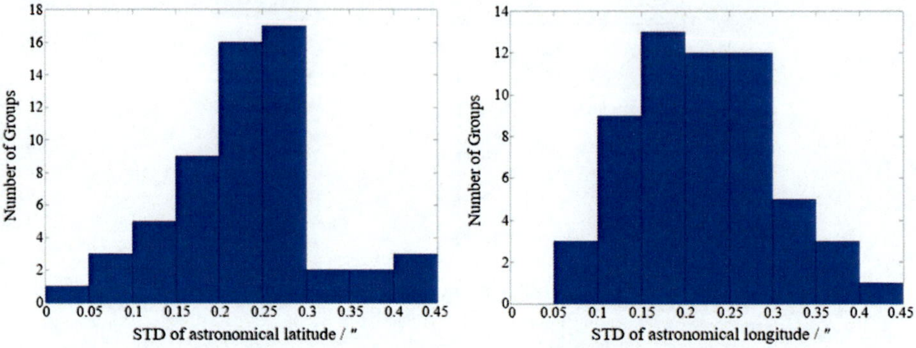

Fig. 4. STD of each observed group from Zhengzhou to Wuhan

From Fig. 4, we can find that there are 51 groups with the astronomical latitude STD less than 0.3″ and 49 groups with the astronomical longitude STD less than 0.3″ among the total 58 groups in the field test. Several groups have STDs of more than 0.3″ under contaminations of observing environment and limited observing time.

Two-group observations made continuously on one point are selected to calculate their differences, seeing Fig. 5. In theory the difference should be equal to zero, but the practical difference is not zero under effects of observing errors. In the experiment, the RMSs of astronomical latitude and longitude from these differences are ±0.074″ and ±0.059″, respectively. The results verified the reliability of this method.

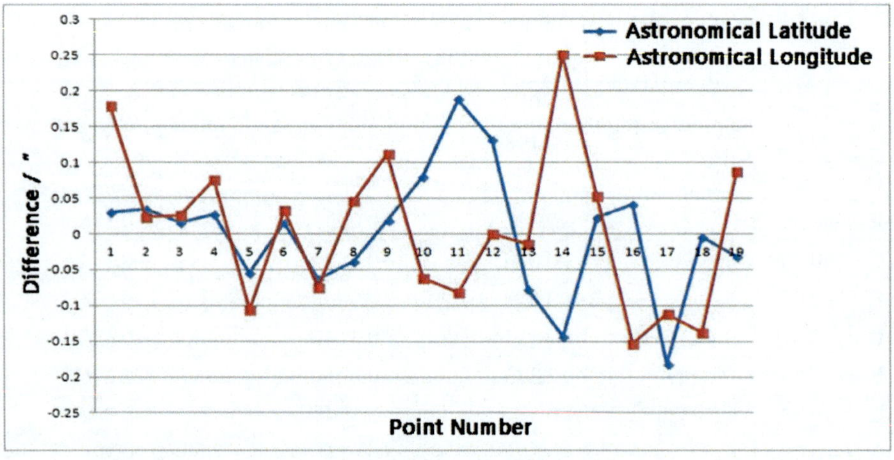

Fig. 5. Differences between two groups of observations of the same point

4.1.4 Observed Results from Beijing to Shenzhen

The field experiments were carried out on 22 feature points in Beijing, Zhengzhou, Luoyang, Sanmenxia, Xianyang, Qinling Mountains, Mount Lu, Nanchang, Gutian and Shenzhen from November 13 to November 28, 2014. 56 groups of observations were captured in the test with the mean observing time of 15 min for each group and the STD of each observed group is shown in Fig. 6.

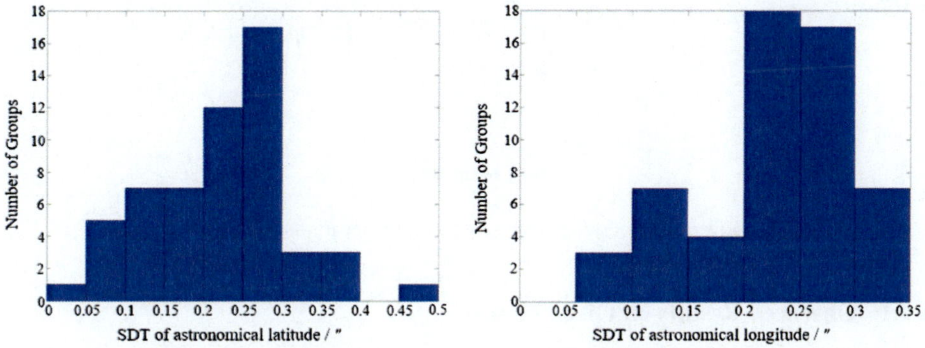

Fig. 6. STD of each observed group from Beijing to Shenzhen

From Fig. 6, we can find that there are 49 groups with the astronomical latitude STD less than $0.3''$ and 49 groups with the astronomical longitude STD less than $0.3''$ among the total 56 groups in the test.

The test results indicate that the observations can be up to preferable precision and the digital zenith tube has a good internal precision. In the field test, the travel distance is larger than 20000 km. The digital zenith tube shows the characteristics of high accuracy and good stability even though it has been in a state of automobile transportation and frequent disassembly.

The differences between two groups of observations of the same point are calculated, seeing in Fig. 7. In the experiment, the RMSs of astronomical latitude and longitude from these differences are $\pm 0.087''$ and $\pm 0.101''$, respectively. This test covered several east and south provinces in China and also passed through several undulating mountains where the plumb line directions change obviously, like Qinling Mountains and Mount Lu. This experiment has important significance to the study of China's high resolution and high precision Earth's gravity field and its time-variation.

4.2 Verification of External Precision

Experiments were also carried out on some known astronomical points at Beijing, Zhengzhou and Xian. Additionally, the vertical deflections observed from Beijing to Shenzhen were compared with EGM2008-derived results. These methods are used to verify the external precision of the astronomical coordinate measured by the digital zenith tube.

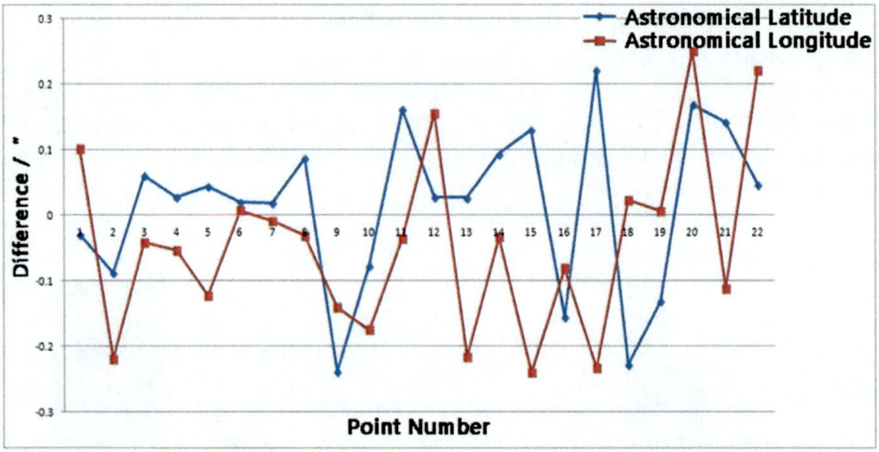

Fig. 7. Differences between two groups of observations of the same point

4.2.1 Observed Results on Given Points

6 groups of observations were captured at Shahe station of Beijing Observatory from September 4 to 9, 2014. The observation time of each group was 15 min. The astronomical coordinate of Shahe had been observed by the traditional astronomical astrolabe for many years. The differences between the given value and observations of Shahe station are depicted in Fig. 8. In the experiment, the mean square errors of astronomical latitude and longitude are $\pm 0.182''$ and $\pm 0.147''$, respectively.

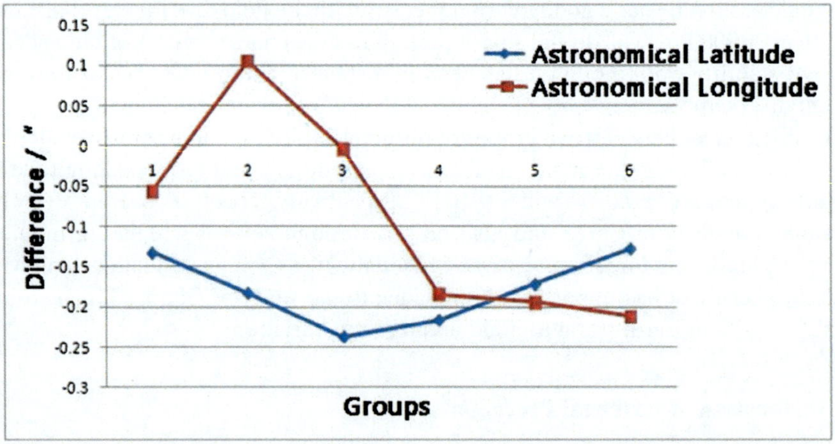

Fig. 8. Difference between given value and observations on Beijing

Field observations were also made on known points in Zhenghou and Xian. The observing astronomical coordinates are compared with the known coordinates determined by the traditional astronomical geodetic method. Table 2 lists the comparing results.

Table 2. Difference between given value and observations of the given point

	Astronomical latitude difference ($''$)	Astronomical longitude difference ($''$)
Beijing	0.128$''$	0.184$''$
Zhengzhou1	0.066$''$	−1.541$''$
Zhengzhou2	0.067$''$	−2.188$''$
Xian	0.005$''$	−1.100$''$

Table 2 shows that the differences of astronomical latitude for all 4 points are less than 0.13$''$. The differences of astronomical longitude in Beijing are less than 0.3$''$, and those of others are greater than 1$''$. The following factors may account for the result.

(1) The known coordinate of Beijing is determined by the long-term observation with one traditional celestial astrolabe, whose principle is similar to that of the digital zenith tube. The others are obtained by the traditional astronomical geodetic method and may include the large personal error, instrumental errors and so on.
(2) The time information adopted by the digital zenith tube is precisely measured by GPS, which can avoid the personal error. However, the time signal utilized in the traditional astronomical geodetic method is broadcasted by the National Time Service of China is received by the astronomical clock, which must deviate from the real time. The abnormal propagation phenomenon of time signal can result in the time delay increasing. In addition, the received time signal needed to record by human, which also causes a time delay. So these time errors seriously affect the accuracy of the astronomical longitude.

4.2.2 The Comparison of Vertical Deflection Between EGM2008 and the Digital Zenith Tube

The Earth gravity model EGM2008 was released by National Geospatial-Intelligence Agency (NGA) with and order of 2159 and degree up to 2190 (Pavlis et al. 2012). EGM2008 is frequently used to calculate height anomaly, geoid undulation, vertical deflection and gravity anomaly (Zhang et al. 2010; Hirt et al. 2013; Malys et al. 2015).

The accuracy of the vertical deflection calculated with EGM2008 in the eastern China is higher than that in the western due to the subdued topography and more gravity data. The comparison of the vertical deflections observed from Beijing to Shenzhen to the EGM08-derived values can also reflect the external precision of the astronomical coordinate measured by the digital zenith tube. The results are listed in Table 3.

Mean differences for meridional and prime vertical components with respect to the EGM08-derived DOVs are −0.302$''$ and 0.020$''$ with the RMSs of 0.559$''$ and 0.480$''$,

Table 3. Comparison of EGM2008 (to degree 2190) vertical deviations and observed vertical deviations

Stations	$\xi_{EGM08} - \zeta_{DZT}/('')$	$\eta_{EGM08} - \eta_{DZT}/('')$
Zhengzhou	−0.405	0.035
Luolong	−0.786	0.526
Sanmenxia	−1.085	−0.437
Xian	−0.059	−0.526
Huangguan	−1.014	−0.44
Qinling	−1.206	−0.199
Fengyukou	−0.071	0.911
Lantian	−0.579	0.026
Shangluo	−0.017	0.166
Lushan	0.206	0.671
Nancheng	−0.693	−0.219
Dabugang	0.028	−0.148
Xiandian	−0.08	−0.338
Nanya	−0.631	0.958
Gongchuan	0.45	−0.158
Gutian	0.085	−0.412
Zhaojiabu	0.094	0.651
Dongshan Island	−0.04	−0.059
Shenzhen	0.069	−0.601

respectively. Although the vertical deflections calculated with EGM2008 are not very accurate, the results indicate the high external precision of the astronomical coordinate measured by the digital zenith tube.

5 Conclusions

One prototype of digital zenith tube integrating GPS and CCD photogrammetry techniques has been developed in China. The principle and error analysis of the digital zenith tube prove that the astronomical coordinate measuring method has the merits of easy operation, time saving, mobile observation and high precise by comparing the traditional astrogeodetic method.

The accuracy of the digital zenith tube was verified by a large number of experiments. In terms of the internal precision, the STD of each group observation was less than $0.3''$ and the RMS of all observations was within $\pm 0.15''$ under good observation conditions. The differences between given values and observed values of the four known points shown the external precision. The differences of all points in astronomical latitude were within ±0.15″. However, the differences of other three points in astronomical longitude are greater than 1″ except Beijing where is less than 0.3″. The possible causes of this case were discussed in the paper. Additionally, the comparison of the vertical deflections observed from Beijing to Shenzhen with the model values of EGM2008 partly reflected the high external precision of the astronomical coordinate measured by the digital zenith tube.

Acknowledgements. This study is supported by the National Natural Science Foundation of China (grant No. 41374009 and 41774001), the Special Project of Basic Science and Technology of China (grant No. 2015FY310200), and the SDUST Research Fund (grant No. 2014TDJH101).

References

Abdujabbarov, A., Amir, M., Ahmedov, B., Ghosh, S.G.: Shadow of rotating regular black holes. Phys. Rev. D Part. **93**, 104004 (2016)

Abele, M., Balodis, J., Janpaule, I., Lasmane, I., Rubans, A., Zariņš, A.: Digital zenith camera for vertical deflection determination. Geodesy Cartograp. **38**, 123–129 (2012)

Bizouard, C., Gambis, D.: The combined solution C04 for earth orientation parameters consistent with International Terrestrial Reference Frame 2008. IERS Earth Orientation, IERS Notice (2011)

Cambiotti, G., Wang, X., Sabadini, R., Yuen, D.A.: Residual polar motion caused by coseismic and interseismic deformations from 1900 to present. Geophys. J. Int. **205**, 1165–1179 (2016)

Dai, J., Guo, J.Y., Jiang, A.H., Yu, X.M., Wang, B.: Denoising of CCD star images measured by automatic measuring system of vertical deflection. Bull. Survey. Map. **10**, 30–33 (2015)

Gaivoronskii, S.V., Berkovich, S.B., Kotov, N.I., Makhaev, A.Y., Sadekov, R.N., Tsodokova, V.V.: An automatic system for determining astronomical Azimuth. Meas. Tech. **58**(3), 280–285 (2015). https://doi.org/10.1007/s11018-015-0700-8

Gerstbach, G., Pichler, H.: A small CCD zenith camera (ZC-G1)-developed for rapid geoid monitoring in difficult projects. Publ. Lobservat. Astron. Beograd **75**, 221–228 (2003)

Guo, J.Y., Song, L.Y., Chang, X.T., Liu, X.: Vertical deflection measure with digital zenith camera and accuracy analysis. Geomat. Inf. Sci. Wuhan Univ. **36**, 1085–1088 (2011)

Guo, J.Y., Song, L.Y., Liu, X., Lu, X.S., Yang, F.L.: Improved two-dimensional moment method of positioning sub-pixel of CCD star image from digital zenith camera. Acta Geodaet. Cartograph. Sinica **40**, 679–683 (2011)

Halicioglu, K., Deniz, R., Ozener, H.: Digital zenith camera system for Astro-Geodetic applications in Turkey. J. Geodesy Geoinf. **1**(2), 115–120 (2012)

Hanada, H., et al.: Development of a digital zenith telescope for advanced astrometry. Sci. China Phys. Mechan. Astron. **55**, 723–732 (2012)

Hauk, M., Hirt, C., Ackermann, C.: Experiences with the QDaedalus system for astrogeodetic determination of deflections of the vertical. Surv. Rev. **49**(355), 294–301 (2017)

Hirt, C.: Monitoring and analysis of anomalous refraction using a digital zenith camera system. Astron. Astrophys. **459**, 283–290 (2006)

Hirt, C., Seeber, G.: Accuracy analysis of vertical deflection data observed with the Hannover digital zenith camera system TZK2-D. J. Geodesy **82**, 347–356 (2008)

Hirt, C., Flury, J.: Astronomical-topographic levelling using high-precision astrogeodetic vertical deflections and digital terrain model data. J. Geodesy **82**, 231–248 (2008)

Hirt, C.: Prediction of vertical deflections from high-degree spherical harmonic synthesis and residual terrain model data. J. Geodesy **84**, 179–190 (2010)

Hirt, C., Bürki, B., Somieski, A., Seeber, G.: Modern determination of vertical deflections using digital zenith cameras. J. Surv. Eng. **136**, 1–12 (2010)

Hirt, C., Claessens, S., Fecher, T., Kuhn, M., Pail, R., Rexer, M.: New ultrahigh-resolution picture of earth's gravity field. Geophys. Res. Lett. **40**, 4279–4283 (2013)

Kong, X.Y., Guo, J.M., Liu, Z.Q.: Foundation of Geodesy, 2nd edn., Wuhan University Press, Wuhan, p. 409 (2010)

Kühtreiber, N.: Combining gravity anomalies and deflections of the vertical for a precise Austrian geoid. Bolletino Di Geofisica Teorica Ed Applicata **40**, 545–553 (1999)

Malys, S., Seago, J.H., Pavlis, N.K., Seidelmann, P.K., Kaplan, G.H.: Why the Greenwich meridian moved. J. Geodesy **89**(12), 1263–1272 (2015). https://doi.org/10.1007/s00190-015-0844-y

Pavlis, N.K., Holmes, S.A., Kenyon, S., Factor, J.K.: The development and evaluation of the earth gravitational model 2008. J. Geophys. Res. **117**, B04406 (2012)

Petrova, N.K., Hanada, H.: Computer simulation of observations of stars from the moon using the polar zenith telescope of the Japanese project ILOM. Sol. Syst. Res. **47**(6), 463–476 (2013). https://doi.org/10.1134/S0038094613060051

Song, G.Y., Wang, J.Y., Cao, Z.C., Song, X.B., Yang, T.L., Zhang, C.H.: The cause of polar movement and its characteristics. Prog. Geophys. **21**, 416–425 (2006)

Song, L.Y.: Research on theory and algorithm for measuring of vertical deflections based on CCD and GPS. Master Thesis, Shandong University of Science and Technology, Qingdao, 81 p (2012)

Tian, L., et al.: Digital zenith telescope prototype of China. Chin. Sci. Bull. **59**(17), 1978–1983 (2014). https://doi.org/10.1007/s11434-014-0256-z

Vittuari, L., et al.: Comparative study of the applied methods for estimating deflection of the vertical in terrestrial geodetic measurements. Sensors **16**, 565 (2016)

Wang, B.: Image and data processing of digital zenith telescope (DZT-1) of China. Ph.D. Thesis, National Astronomical Observatories of Chinese Academy of Sciences, Beijing, 67 p. (2014)

Wang, B., et al.: Image and data processing of digital zenith telescope (DZT-1) of China. Chin. Sci. Bull. **59**(17), 1984–1991 (2014). https://doi.org/10.1007/s11434-014-0277-7

Zariņš, A., Rubans, A., Silabriedis, G.: Digital zenith camera of the University of Latvia. Geodesy Cartograp. **42**(4), 129–135 (2016)

Zhang, J.M., Yan, J.Q., Wang, F.M.: Accuracy analysis and evaluation of EGM2008 Earth gravity field model. Oil Geophy. Prospect. **45**, 230–233 (2010)

Properties of Low Strength and High Fluidity Recycled Aggregates

Wen Zhao[1,2,3], Yingbiao Wu[1,2,3], Jinjin Shi[1,2,3](✉), and Jinyan Liu[1,2,3]

[1] Cangzhou Municipal Engineering Company Limited, Cangzhou 061000, China
[2] Hebei Province Road Materials and Technology Engineering Technology Research Center, Cangzhou 061000, China
[3] Hebei Industrial Technology Research Institute of Construction Waste Comprehensive Utilization, Cangzhou 061000, China

Abstract. Construction waste was processed into recycled coarse aggregates of 5–20 mm and fine aggregates of 0–5 mm by using a mobile crushing and screening process. Cementitious materials such as cement, fly ash, asphalt mixture reclaimed dust, and additives were added to prepare the low strength and high fluidity materials. Volumetric method was used to design the mix proportion of different materials. The optimum mix proportion of 0–5 mm recycled fine aggregates and 5–20 mm recycled coarse aggregates for the mix design was found to be 33.5% and 66.5%, respectively. The workability and mechanical properties of low strength and high fluidity materials mixed with different cementitious materials were investigated. The results show that the material workability decreased with increasing asphalt mixture reclaimed dust and increased after adding water reducing admixture. The 28-day compressive strength is found to be between 1.4 and 4.1 MPa meeting the material specification requirements.

1 Introduction

With the continuous acceleration of China's urbanization, the number of excavation backfill projects in various regions is increasing. The traditional backfill construction generally adopts graded sand, gravel packing, or excavation soil to meet the specification requirements for layered filling and compaction. However, due to the narrow construction site and the existence of underground structures, it is often impossible to use conventional compactors and other compaction tools that lead to difficulties in ensuring the compaction quality and often induces engineering distresses [1, 2]. On the other hand, a large number of construction wastes are generated in the process of demolition and reconstruction. Many cities and regions have seen an increasing phenomenon of 'encircle the city with garbage'. Therefore, it becomes a crucial venture to make full utilization of resources that are attribute of construction wastes.

Low strength and high fluidity materials are generally prepared by utilizing the recycled aggregates of the construction waste as raw materials. With their high fluidity, they can be filled to form a self-compacting structure under the action of dead weight without or a little vibration [2, 3]. They can also be constructed in narrow and hard-to-reach places. Due to their low strength and relatively low requirements for raw materials

A. Tapase et al. (Eds.): GeoChina 2021, SUCI, pp. 47–56, 2021.
https://doi.org/10.1007/978-3-030-79644-0_5

[4], the recycled aggregates are used to replace the natural stones and the reclaimed dust to replace a part of fly ash. On the premise of ensuring the mechanical properties of materials, it not only saves natural raw materials, but also provides an effective way for the resource utilization of construction waste.

2 Regeneration Treatment Process of Construction Waste

The mobile construction waste treatment process is adopted to recycle brick and concrete construction waste from Cangzhou municipal material yard. The treatment process mainly includes pretreatment, sorting and removal of impurities, crushing, and screening [5] (Fig. 1).

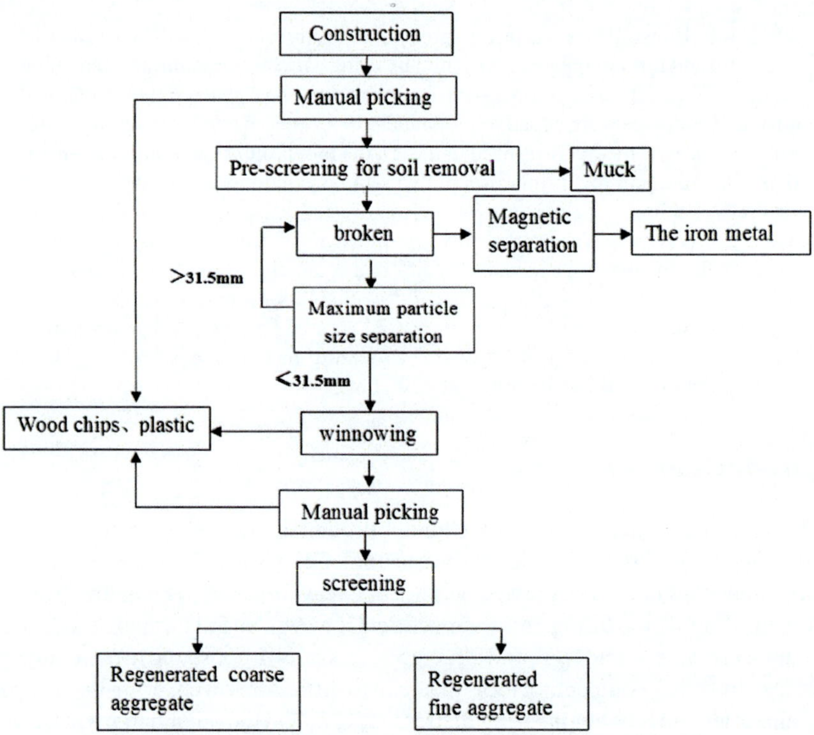

Fig. 1. Mobile construction waste treatment process

Among them, the crushing task mainly depends on the mobile crush equipment. Mobile crush equipment is comprised of vibrating feeder, crusher, screening system, and sundry sorting device. Transmission mechanism is the integration of the combination equipment. The recycling of the construction waste was first broken by mechanical means, and then separation of steel and iron was processed by magnetic method. The aggregates larger than 31.5 mm were separated by the super-size screening for the secondary crushing. The resulting materials less than 31.5 mm, went through continuous gradation of the recycled aggregates, were kept in storage for the next process.

The screening task usually relies on the mobile screening machine. Two layers of screen were set according to the classification requirements, which were 5 mm and 20 mm size screen. The broken recycled aggregates were screened out to three types aggregates (0–5 mm, 5–20 mm and 20–31.5 mm). The recycled aggregate of 0–5 mm and 5–20 mm were used to prepare the low-strength and high-fluidity materials.

3 Raw Materials and Test Methods

3.1 The Raw Material

(1) Recycled coarse aggregate (5–20 mm)
The recycled coarse aggregates were produced by crushing and screening in the Cangzhou municipal material yard. The test results of various physical performance indices are shown in Table 1.

Table 1. Physical performance test results of reclaimed coarse aggregates

Test items	Bulk density (kg/m^3)	Compact density (kg/m^3)	Apparent density (kg/m^3)	Porosity (%)	Water absorption(%)
Test results	1080	1290	1420	23.8	9.2

(2) Recycled fine aggregate (0–5 mm)
The physical performance indices of the recycled fine aggregates are shown in Table 2.

Table 2. Physical properties of reclaimed fine aggregate

Test items	Bulk density(kg/m^3)	Compact density (kg/m^3)	Apparent density (kg/m^3)	Porosity (%)	Water absorption (%)
Test results	1150	1470	2110	45.5	31.9

(3) Recycled fines
In the production process of asphalt mixture, a large amount of dust is collected due to heating and dust removal of aggregates, which is mainly a mixture of mineral powder and impurities, called recycled fines. If all the recycled fines are discarded, it would cause a great waste and environmental pollution.

The recycled fines were produced by the mixing equipment of the Cangzhou Municipal asphalt mixing station in the production process. The recycled fines were mainly from the fine aggregate mechanism sand used to produce the asphalt mixture. The test results of various physical performance indicators are shown in Table 3.

Table 3. Physical performance test results of recycled fine

Test items	Particle size range			density (g/cm^3)	Specific surface area (m^2/kg)	Coefficient of hydrophilic	Liquid limit (%)	Plastic limit (%)	Plastic index
	<0.6 mm	<0.15 mm	<0.075 mm						
Test results	100	99	98.5	2.68	671	0.52	36.1	26.1	10

(4) Cement
The cement used in this study was Jidong "Shield stone" P.O42.5 cement. After testing, its performance indices, as shown in Table 4, met the requirements of "General Portland Cement" (GB 175-2007) [6].

Table 4. Test results of performance indicators of cement

The test items	Specific surface area (m^2/kg)	Setting time (min)		Flexural strength (MPa)		Compressive strength (MPa)		Ignition loss (%)	Stability
		Initial setting	Final setting	3-day	28-day	3-day	28-Day		
Test results	375	145	258	5.8	8.7	28.4	50.5	4.2	Qualified
Technical requirem-ents	≥300	≥45	≤600	≥3.5	≥6.5	≥17.0	≥42.5	≤5.0	Qualified

(5) Fly ash
The fly ash were from the China's resources power plant with F-class II grade. After examination, its performance indicators (Table 5), were found to meet the requirements of fly ash used in Cement and Concrete (GB/T 1596-2017) [7].

Table 5. Test results of fly ash performance indexes

Test items	Fineness (45 μm squared sieve residue) (%)	Specific surface area (m^2/kg)	Density (g/cm^3)	Water demand ratio (%)	Ignition loss (%)	Water content (%)	Intensity activity index (%)
Test results	23.1	317	2.1	102	1.2	0.4	75
Technical requirements	≤30.0	-	≤2.6	≤105	≤8.0	≤1.0	≥70.0

(6) Admixture
The admixture was a PA-I polycarboxylic acid superplasticizer. After testing, the concrete performance indices of the admixture met the requirements of "Concrete admixtures" (GB 8076–2008) [8], as shown in Table 6.

Table 6. Test results of concrete performance indexes of admixture

Test items	Water reducing rate(%)	Ratio of bleeding rate(%)	Air content(%)	Compressive strength ratio(%)	
				7d	28d
Test results	18	75	3	150	135
Technical requirements	≥14	≤100	≤4.5	≥125	≥120

3.2 Test Methods

(1) Workability
The workability of low strength and high fluidity materials is mainly controlled by slump and expansion, liquidity of auxiliary observation concrete, adhesiveness, and water retention. In order to evaluate the workability of concrete, the slump and expansion were tested following the Standard for Performance Test Methods of Common Concrete Mixtures (GB/T 50080-2016) [9].

(2) Mechanical properties
The mechanical properties of low-strength and high fluidity materials are characterized by the 7-day and 28-day compressive strength tests. The tests were conducted in accordance with the compressive test method specified in the Standard for Testing Methods for Mechanical Properties of Common Concrete (GB/T 50081-2002) [10].

3.3 Test Mix Proportion

Due to the large variability of the recycled aggregates, the volumetric method was selected to design a benchmark mix ratio for the low strength and high fluidity materials. According to the production experience, the following cementitious materials and their contents were determined:

(1) Cement content - 50 kg;
(2) Fly ash content - 140 kg; and
(3) Recovered powder content-140 kg.

The amount of 0–5 mm and 5– 20 mm recycled aggregates was calculated based on Eq. 1 and Eq. 2, respectively.

$$S = \rho_S \times P/(1 - H_s) \tag{1}$$

Where: S—0–5 mm The amount of aggregate, kg;

ρ_S—0–5 mm Aggregate compact density, kg/m^3;

P—5–20 mm Void fraction of aggregate, %; and

H_s—0–5 mm

The proportion of powder finer than 0.075 mm was 6.1%, which was used as a cementing material.

$$G = (1 - V_1 - V_2 - V_3 - V_s - V_w - P) \times \rho_{gapparent} \tag{2}$$

Where: G—the amount of aggregate (5–20 mm), kg

V_1, V_2, V_3, V_s, V_W—the volume of mixing water for cement (m^3), fly ash, recovered fly, powder less than 0.075 mm in 0–5 mm aggregates, and cementitious material, respectively;

$\rho_{gapparent}$ is the apparent density of aggregate (5–20 mm),kg/m^3。

The optimal mixing ratios of 0–5 mm and 5–20 mm recycled aggregates were found to be 33.5% and 66.5%, respectively. The water consumption of the mixture was computed according to the water demand of each cementing material and the water absorption rate of aggregates. Considering the calculated optimal mixing ratios of aggregates, material's economy, different dosages of cementing material, and concentration of the selected admixture, a total of eight different groups of mix proportions, as illustrated in Table 7, were designed for the subsequent testing.

Table 7. Mix proportions of low strength and high fluidity materials

Mix proportion number	Cement (kg)	Fly ash (kg)	Recycled Fly (kg)	Recycled aggregate 0–5 mm (kg)	Recycled aggregate 5–20 mm (kg)	Water (kg)	Admixture (kg)
1	50	140	140	372	739	318	0
2	50	0	280	372	739	318	0
3	90	120	120	372	739	318	0
4	90	0	240	372	739	318	0
5	50	140	140	372	739	278	12
6	50	0	280	372	739	278	12
7	90	120	120	372	739	278	12
8	90	0	240	372	739	278	12

4 Test Results and Analyses

4.1 Workability

(1) Influence of cementing materials on workability

For the mix proportions #1–#4, the quantity of the recycled fine and coarse aggregates were kept the same. The proportions of cement, fly ash, and the recovered powder less than 0.075 mm were varied (Table 7). The results of the performance tests are presented in Table 8, Fig. 2, and Fig. 3. It can be seen from Table 8, the fluidity,water-holding capacity, and cohesiveness of the mixture are good. As shown in Figs. 2 and 3, the slumps of the mixtures are between 200–220 mm, and the expansions are between 330–350 mm.

This indicates that with increasing the fly ash content, the mixture's slump and expansion increased. That leads to better workability or working performance. This is because fly ash is composed of spherical vitreous of different sizes, whose surface is smooth and dense, and particle size is smaller than cement. It leads to better lubricating effect in the mixture. The fly ash also has obvious morphological effects, which can improve the workability or working performance of materials.

The reclaimed dusts are the small dust recovered in the asphalt mixture production process. Its particle shape is irregular, and the specific surface area is larger than fly ash. The test results show that the addition of the reclaimed dust had a negative impact on the workability of the material. When the mix proportion # 2 ratio of the reclaimed dust content is the highest, the slump and expansion ratios are the lowest.

Table 8. Performance test results of mixtures

Mix proportion number	Liquidity	Water retention	Adhesiveness
1	General	Good	Good
2	General	Good	Good
3	General	Good	Good
4	General	Good	Good

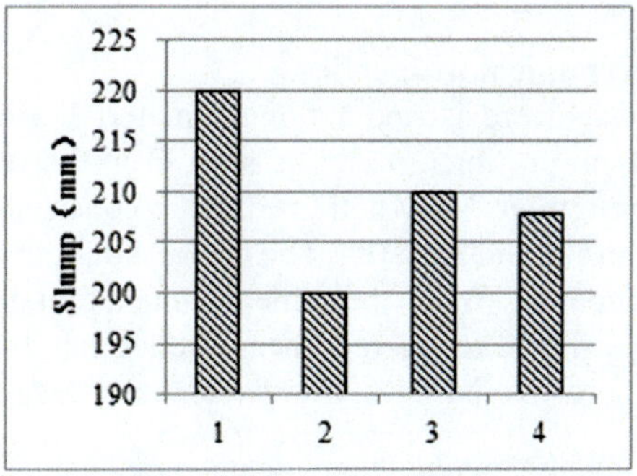

Fig. 2. Slump test results

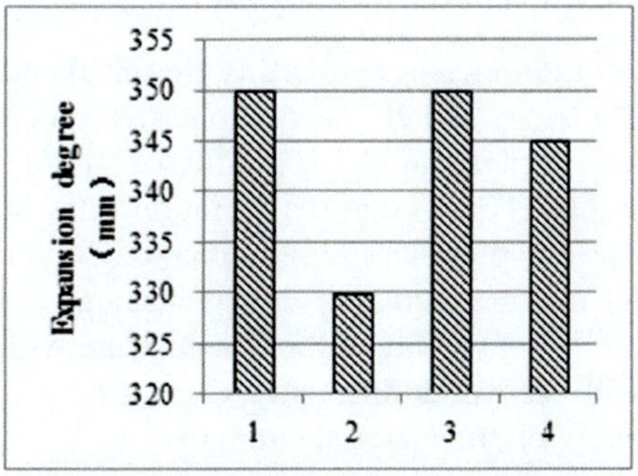

Fig. 3. Expansion degree test results

(2) Effects of additives on workability

Low strength and high fluidity materials require good fluidity. Therefore, an equal amount of polycarboxylic acid, a highly efficient water reducing admixture,was added to produce the mix proportions #5–#8 (Table 7). These mix proportions were used for the slump and expansion detection. As can be seen from Table 9, these mixtures with admixture have a good fluidity. As shown in Figs. 4 and 5, after adding admixture, the material's slump and expansion have been significantly improved with the maximum slump of 130 mm and the maximum expansion of 350 mm. The change trend is consistent with the respective mix proportion and no admixture.

Table 9. Operating performance test results of mixtures

Mix proportion number	Liquidity	Water retention	Adhesiveness
5	Good	Good	Good
6	Good	Good	Good
7	Good	Good	Good
8	Good	Good	Good

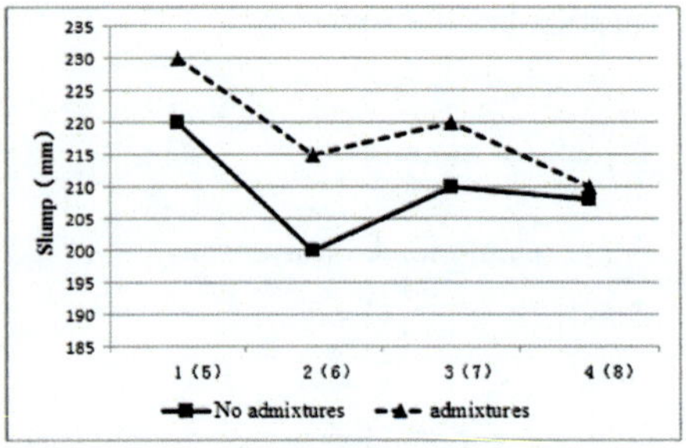

Fig. 4. Slump test results

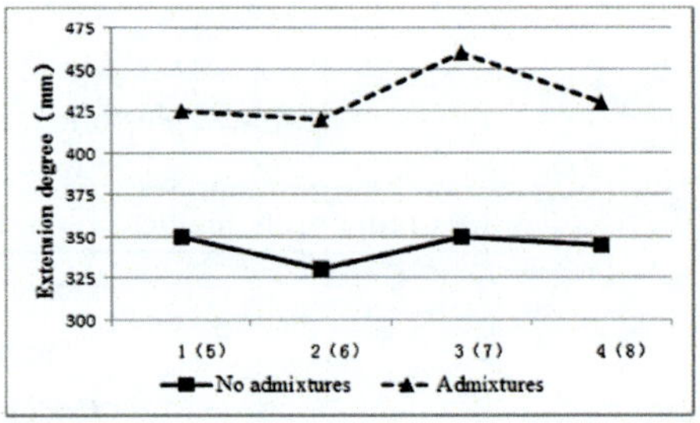

Fig. 5. Expansion degree test results

4.2 Mechanical Properties

(1) Influence of cementing materials on mechanical properties

The 7-day and 28-day compressive strength tests were carried out on the materials with low strength and high fluidity of #1–#4 mix proportions. It can be seen from the test results (Fig. 6) that the variation rules of strength at 7-day and 28-day are consistent. The compressive strength of 28-day is between 1.4 and 3.3 MPa. The overall strength of the mixture increases with increasing cement content. When the cement content remains the same, the strength of the mixture with fly ash and recovered fines is better than that with recovered fines only. Compared with the recovered fines, the fly ash has a greater effect on the material strength.

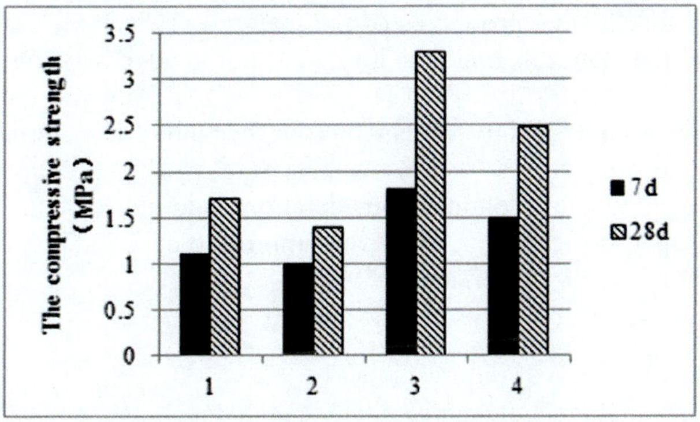

Fig. 6. Compressive strength test result

(2) Effects of additives on mechanical properties
As shown in Figs. 7 and 8, the compressive strength of the materials with low strength and high fluidity of #5–#8 mix proportions have been tested for 7 and 28 days. It can be seen that, after adding the additive, the compressive strengths of materials have significantly been improved.

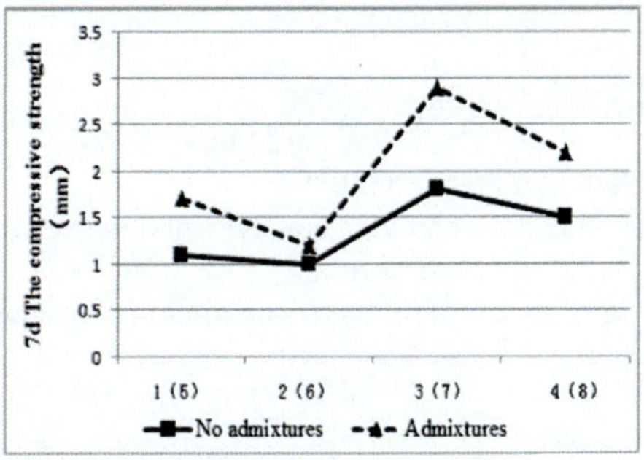

Fig. 7. Comparison of compressive strength test results on 7-day

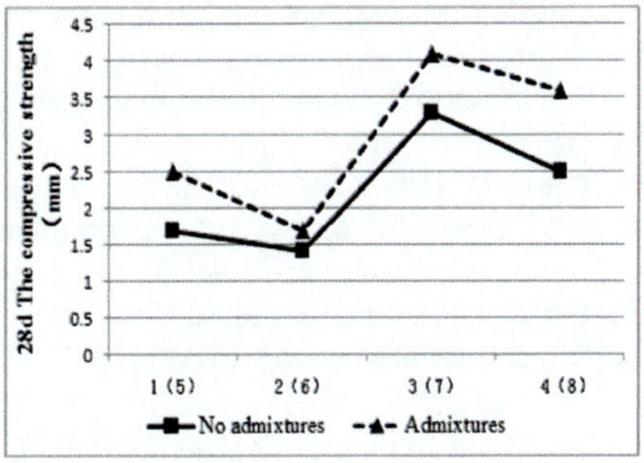

Fig. 8. Comparison of compressive strength test results on 28-day

5 Conclusions

The following conclusions and observations are adrawn:

(1) The low strength and high fluidity materials that meet the requirements can be prepared by adding appropriate amount of cementation materials into the recycled coarse and fine aggregates of brick and concrete construction waste. The slump was found to be between 200 and 230 mm and the expansion was in between 330 and 460 mm with good fluidity. The compressive strength at 28-day was between 1.4 and 4.1 MPa.

(2) For the low strength and high fluidity materials, a certain amount of reclaimed dust can replace fly ash reducing the production cost and providing a way for recycling the reclaimed dust. The addition of the reclaimed dust had a certain adverse effect on

the working and mechanical properties of the material. Therefore, the amount of the reclaimed dust in the material should be limited as not to adversely impact its liquidity and strength.

(3) Increasing the content of fly ash can improve the fluidity of materials by increasing its slump and expansion.

(4) Under the premise of economic considerations, adding a certain amount of polycarboxylic acid superplasticizer can effectively improve the workability and enhance the compressive strength of the material.

References

1. Meng, L., Li, Z., Fei, L.: Experimental study on the preparation of controllable low-strength materials from construction abandoned soil. Jiangxi Build. Mater. **12**, 56–60 (2015)
2. Jinxi, Z., Ting, L., Bo, S., Jiangang, W., Bo, W.: Study on the performance of bentonite containing construction waste flowing backfill material. Municipal Technol. **35**(06), 173–175 (2017)
3. Shugang, W., Hong, Z., Zhiyuan, W.: Research on the properties of controllable low-strength backfill materials. Inner Mongolia High. Transp. **04**, 10–13 (2012). (in Chinese)
4. Fei, L., Chenhui, L., Yingbiao, W., Fujin, W., Li'an, Z.: Influence of recycled construction waste materials on the performance of controllable low strength materials (CLSM). Concrete **08**, 71–73+78 (2018)
5. Yingbiao, W., Jinjin, S., Jinyan, L., Wen, Z.: Comprehensive application of construction waste in urban road engineering. Constr. Technol. **23**, 33–36 (2016)
6. GB 175-2007: General Portland cement. China Standard Press (2007)
7. GB/T 1596-2017: Fly ash used in cement and concrete. China Standard Press (2017)
8. GB 8076-2008: Concrete admixtures. China Standard Press (2009)
9. GB/T 50080-2016: Performance test method standard for common concrete mixtures. China Building Industry Press (2017)
10. GB/T 50081-2002: Standard of test methods for mechanical properties of common concrete. China Building Industry Press (2003)

Influence of Traffic Characterization Methodology on Service Life Prediction of Pavements Subjected to Overweight Traffic Operations

Ali Morovatdar(✉) and Reza S. Ashtiani

Department of Civil Engineering, The University of Texas at El Paso, 500 W. University Avenue, El Paso, TX 79968, USA
amorovatdar@miners.utep.edu, reza@utep.edu

Abstract. Mechanistic-Empirical (ME) design and analysis of pavement structures are highly sensitive to the traffic data inputs. Hence, proper characterization of truck traffic is the key step for accurate assessment of performance and service life of pavements subjected to overweight (OW) traffic movements. Therefore, this study aimed to evaluate the impact of different traffic characterization methodologies on the predicted service life of pavements in overload corridors. To accomplish this objective, initially, the authors collected the site-specific traffic information by deploying Portable Weight-in-Motion (P-WIM) devices to ten representative pavement sections in Texas. Subsequently, non-destructive field tests such as Falling Weight Deflectometer (FWD) and Ground Penetrating Radar (GPR) were performed during summer and winter months for the back-calculation of the layer moduli in representative sites. Ultimately, different traffic data inputs, including the site-specific Axle Load Spectra (ALS), Equivalent Single Axle Load (ESAL) values, and software default values, as well as the pavement layer properties were incorporated into a series of ME analyses to estimate the service life of the pavements. The comparative analysis results indicated that the incorporation of ESAL and software default values in the analysis, instead of using site-specific ALS, substantially overestimated the service life of the pavements. Such deviation was more pronounced for pavement sections of Farm-to-Market roadways subjected to unprecedented OW traffic operations. Therefore, the traditional traffic characterization methodologies were not capable of representing the demanding loading conditions in the studied network. Further comparisons between the post-processed results and field distress records in this study showed that the most promising methodology to incorporate the traffic information into the service life analysis of pavements in overload corridors is the concept of the site-specific ALS.

Keywords: Mechanistic-Empirical (ME) analysis · Pavement service life · Axle load spectra · Portable Weight-in-Motion · Texas overload corridors · Over-Weight (OW) trucks

A. Tapase et al. (Eds.): GeoChina 2021, SUCI, pp. 57–74, 2021.
https://doi.org/10.1007/978-3-030-79644-0_6

1 Introduction

In the past decades, several states have experienced drastic improvements in energy-related activities such as natural gas and crude oil production, as well as increased freight transportation due to improvements in the economy of States. Energy development operations have provided favorable economic benefits to the states and the nation. However, such activities have significantly increased the frequency of the heavy truck traffic movements in the network, resulting in expedited deterioration of ride quality and loss of pavement service life in the impacted zones. Damaged local and county roads have been a major source of inconvenience for the local residents in such states. A prime example of that is the unprecedented energy production activities during the past decade in highly active oil fields in the Permian Basin and the Eagle Ford Shale regions in Texas. The energy production activities on one hand, and the increase in freight transportation due to improvements in economic activities of the state on the other, have resulted in unprecedented operations of Over-Weight (OW) trucks. The taxing magnitude and frequency of such loading scenarios resulted in substantial damages to the transportation facilities.

Due to the sudden explosion of drilling activities, the local government agencies and department of transportations across the nation were not able to ramp up their pavement preservation and maintenance efforts to restore the ride quality of the damaged pavements. Lack of funding resources coupled with unclear guidelines are among the many elements that contribute to the delay of the pavement maintenance and rehabilitation in several counties across the nation. Accurate quantification of the structural impacts of OW trucks is the prelude to adopting proper rehabilitation strategy by stakeholders. This can effectively protect the taxpayers' resources allocated to restore the current transportation facilities and expand the network to meet future traffic demands. Consequently, there is a pressing need to accurately quantify the pavement loss imparted by OW truck operations in the affected networks.

An accurate account of the traffic characteristics is the precursor for reliable prediction of the service life of the pavements subjected to OW traffic loads. Traditionally, the analysis of service life is based on the Equivalent Single Axle Load (ESAL) concept, in which the traffic mix is converted to 18-kips standard single axle. Commonly, the ESAL values are calculated using the empirical AASHTO formulations with several simplifying assumptions. The root of the problem that pavement practitioners face is that the analysis procedures merely rely on experimental information that was developed from field measurements in the late 1950s and early 1960s, with revisions later in 1993. The ESAL concept also overlooks the influence of variations in the material properties throughout the year and with incurring damages during the service life of the pavements. The assumptions made in this approach can potentially induce systematic error for the calculation of the pavement damages and distresses during the design life of the pavements. To mitigate this anomaly, new mechanistic-empirical (ME) design guides, such as the Texas Mechanistic-Empirical (TxME) pavement design system, incorporate incremental damage concept, in combination with axle load spectra (ALS) to realistically simulate the progression of the distresses with time. This will allow for the determination of incremental damage imparted by a specific vehicle class at a specific timeframe on the pavement.

The site-specific ALS needed for ME pavement analysis is mainly collected from Weigh-In-Motion (WIM) units in the field. There are two main categories of WIMs in the pavement industry; Portable WIMs and WIM stations. The Portable WIM (P-WIM) systems are preferred due to their convenience, cost-efficiency, and flexibility for continuous collection of the traffic data without interrupting the traffic flow in heavily trafficked highways. Additionally, the P-WIM units are capable of collecting reliable and accurate traffic data provided that a verified calibration procedure is implemented upon the installation of the piezo-electric sensors in the field. Consequently, several researchers favored P-WIM systems over the traditional stationary WIM devices in order to collect traffic information (Ashtiani et al. 2019; Morovatdar et al. 2020a).

Several researchers studied different methodologies to evaluate the service life of pavements subjected to OW truck traffic. Wang et al. (2015), using WIM traffic data, investigated the impact of OW vehicles on the service life of the pavements in New Jersey. Based on numerical simulations, the authors reported that by increasing OW truck frequencies, the service life of the pavements can be substantially reduced in their studied network. Rys et al. (2016) assessed the influence of overloaded vehicles on fatigue life of pavements based on the WIM data. The authors converted the traffic mix to the equivalent number of standard axles to characterize the traffic data in their analysis of OW zones. Based on analyzed sections in their study, 20% increase in OW truck operations resulted in 50% reduction in the service life of the pavements. Banerjee and Prozzi (2015) developed a methodology for the determination of the consumption of pavement service life due to the OW traffic operations in Texas. The researchers used the ESAL concept and the pavement damage equivalency factors to characterize the traffic mix in their analysis. The authors in turn used this information to propose a permit fee structure for OW trucks. In a relevant study, Batioja-Alvarez et al. (2018), using the ESAL concept, evaluated the loss of pavement service life imparted by the OW truck operations in Nevada. Evidently, they reported that the reduction of pavement life was influenced by the environmental conditions and asphalt layer thicknesses. The researchers used records of issued permits as the primary source to characterize the traffic distributions in the network.

Several researchers used Pavement Condition Score (PCS) and roughness data in lieu of mechanistic-based approaches to assess the performance deterioration and service life reduction of the pavements in overload corridors (Chou et al. 2008; Al-Qadi et al. 2017; Romero et al. 2019). The PCS takes into account both the Distress Score (DS) and the International Roughness Index (IRI) to provide a quantitative means to describe the ride quality and pavement conditions. Though not intended as a mechanistic indicator of the distress progressions in pavements, some researchers have preferred its use as a means to predict the service life of pavements in the absence of traffic distributions and pavement structural data. In some states such as Texas, due to the lack of a uniform approach for mechanistic determination of the loss of pavement service life, prioritization of pavement sections to conduct Maintenance and Rehabilitation (M&R) efforts is traditionally based on the PCS and relevant information gathered from visual inspection surveys.

Recently, several researchers used advanced statistical clustering techniques to categorize pavement sections with similar traffic patterns for further remaining service life analysis (Smith and Diefenderfer 2010; Haider et al. 2011; Mai et al. 2013; Abbas et al.

2014; Oh et al. 2015; Li et al. 2016; Jasim et al. 2019; Walubita et al. 2019). The comparative analysis between different input levels for ALS indicated that the cluster-based ALS can serve as a means for the analysis of the pavement service life in the absence of field ALS database.

The primary objective of this research study was to assess and compare the impact of different traffic input scenarios, i.e., site-specific ALS, ESAL values, and TxME default values, on the predicted performance and service life of pavements subjected to the OW truck operations in Texas overload corridors.

2 Representative Sites for Field Testing

In this study, prominent features associated with the traffic patterns and material properties of pavement layers were directly determined from field data collection and non-destructive testing of ten representative sites strategically distributed throughout the energy developing areas of Texas. Figure 1 shows the locations of the sites selected for field-testing comprised of several Farm to Market (FM) roads, State Highways (SH), and US Highways across the Eagle Ford Shale region.

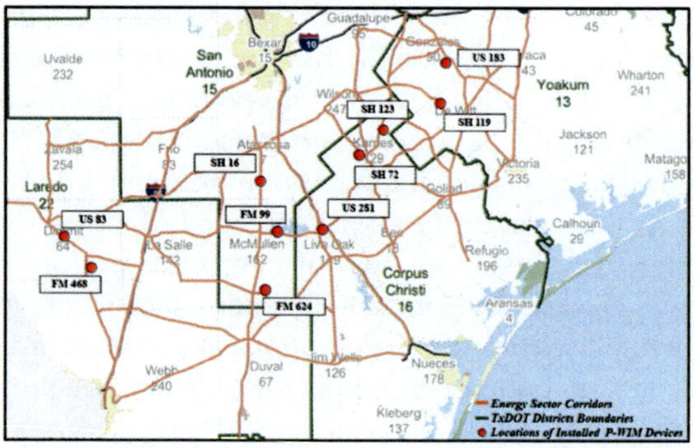

Fig. 1. Selected sites in Texas overload corridors.

3 Prediction of Service Life of Pavements Using Different Traffic Inputs

Figure 2 shows the flowchart of the procedure adopted to incorporate various traffic input data into the analysis for the determination of the service life of pavements. The authors implemented the mechanistic approach developed in their recent study for prediction of the remaining service life of the pavements in overload corridors (Ashtiani et al. 2019). As illustrated in the figure, the mechanistic-empirical pavement design software employed in this study, TxME, requires four main categories of input, i.e., structure, traffic, climate, and performance criteria limits. To characterize the design inputs, initially, the pavement layer properties were obtained from post-processing of the Ground Penetrating Radar (GPR) and Falling Weight Deflectometer (FWD) field data. The pavement design plans were further instrumental in validating or supplementing the information on the material properties, layer configurations, and rehabilitation history of the pavement layers. Subsequently, the authors deployed P-WIM devices to collect site-specific ALS during summer and winter months, for accurate characterization of traffic in ten representative sites in the studied network.

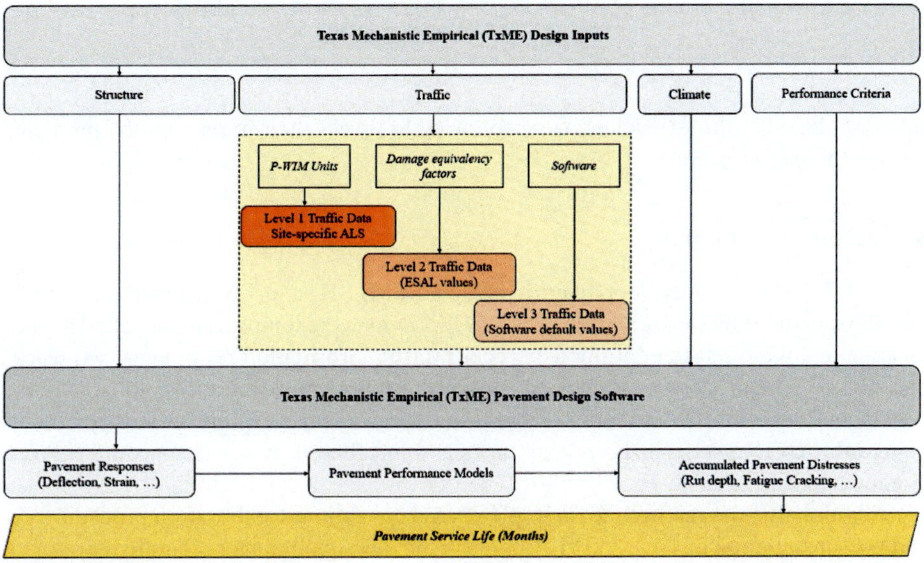

Fig. 2. Flowchart for incorporation of different traffic input scenarios into the ME analysis procedure for the determination of the pavement service life.

The site-specific pavement layers properties and traffic loading information were then incorporated in the TxME pavement design software to obtain pavement responses. Climatic data from an adjacent weather station, as well as performance criteria limits, were also assigned using the agency defined thresholds. Subsequently, the critical pavement responses for multiple axle weights and axle/tire configurations were calculated in this study. Based on the internal algorithms and incorporated performance models for the quantification of pavement damages, the incremental increases in pavement damages were calculated. Ultimately, the service life of pavements were determined based on the number of months until the cumulative pavement distresses meet the agency defined threshold. The pavement performance criteria considered in this study include cumulative rutting at the pavement surface, asphalt concrete (AC) layer fatigue cracking, stabilized base layer fatigue cracking, and thermal cracking. The distress thresholds set forth by TxDOT are shown in Table 1.

Ultimately, the authors conducted a series of numerical simulations in order to highlight the influence of various methods of traffic characterization in the analysis of the service life of pavements. Hence, similar processes but with different traffic input scenarios, including the site-specific ALS (level 1), ESAL values (level 2), and ultimately the

Table 1. TxME pavement performance criteria limits

Performance criteria	Limit
Total surface rutting (inch)	0.5
Thermal cracking (ft/mile)	2112
AC fatigue cracking area (%)	50
Stabilized base fatigue cracking (%)	50

software default values (level 3), were followed to assess their impact on the predicted service life of pavements.

3.1 Level 1 Traffic Data

The P-WIM devices were deployed in selected FM, SH, and US roadways to collect the site-specific traffic information. The P-WIM units were temporarily installed at each location and were left to continuously record traffic information for at least two weeks in each site. The process was repeated for both summer and winter times to capture the seasonal effect of traffic variations. Field installation consisted of piezo-electric sensors being inserted into specialized pocket tapes and installed in two sets spaced 8 ft apart, as shown in Figs. 3a and 3b.

Additionally, the research team implemented an elaborate calibration procedure pre and post-installation of the P-WIM systems in the field to obtain an accurate representation of the traffic data from the P-WIM data (Morovatdar et al. 2020b). Information on the current traffic mix, average annual daily truck traffic, vehicle class distribution, axle configuration, and the axle load spectra, describing Level 1 traffic inputs, was captured by the P-WIM units in the field.

(a) (b)

Fig. 3. Deployment of the P-WIM devices in the field, (a) Typical P-WIM system setup, (b) Arrangement of installed piezo-electric sensors.

As stated earlier in the introduction section, ALS provides the most desirable traffic data input for ME pavement design and analysis. Hence, the traffic data collected by

P-WIM units were synthesized to develop the site-specific ALS database in this study. Then the ALS were analyzed and classified by different axle types to develop axle load distributions within each vehicle class and each individual season of the year.

Figure 4 illustrates the tandem-axle load distributions in one of the heavily trafficked sections, State Highway 123 in Corpus Christi District, based on the traffic data collected by P-WIM devices during the winter of 2019. The analysis of the load distributions in SH123 indicated high number of OW truck operations in this roadway. The plot shows that nearly 44% of all tandem axles were attributed to the passages of OW axles in SH 123 during the studied timeframe. Such high percentages of OW truck traffic in the network can potentially jeopardize the longevity of transportation facilities and result in premature failure of pavement structures. The relevant information on the traffic trends, load distributions, as well as the synthesized findings attributed for all ten representative sites, is provided in Ashtiani et al. (2019).

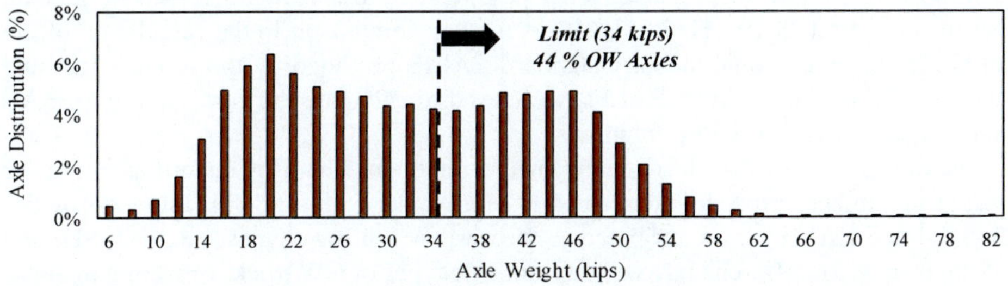

Fig. 4. Tandem axle load distributions for all trucks in State Highway 123, Corpus Christi.

Truck class distributions are essential components of traffic information that significantly influence the analysis of the service life of pavement structures. Based on the traffic data from the P-WIM systems in this study, the most prevalent OW vehicles were Class 5 and Class 9 trucks. Figure 5 illustrates the truck class distributions recorded by P-WIM units in SH 123. As evidenced in the plot, Class 9 trucks with 65% of total truck traffic were the most predominant overweight truck class, followed by Class 5 trucks, with a frequency of 12% of the total truck traffic.

3.2 Level 2 Traffic Data

The projected ESAL values over a 20-year design life, as level 2 traffic inputs, were calculated from Eq. 1 as:

$$ESAL_{current} = \sum_{i=1}^{m} (EALF)_i \, n_i \tag{1}$$

where:

$(EALF)_i$ = *Equivalent Axle Load Factor (EALF),*
n_i = *Number of passes of* i^{th}*-axle load group.*

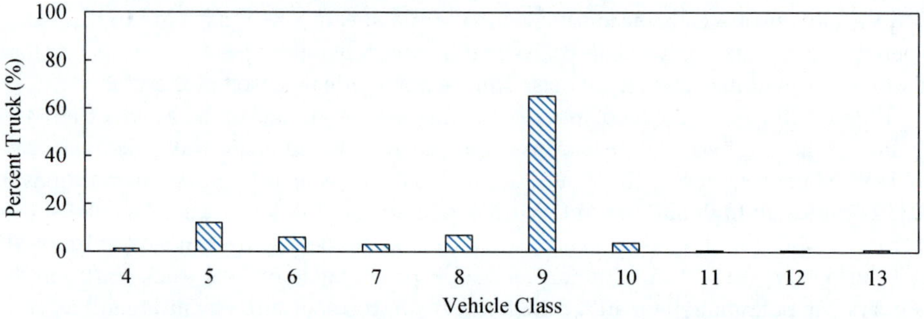

Fig. 5. Recorded truck class distributions in State Highway 123.

The Equivalent Axle Load Factors (EALFs) essentially allow for the quantification of the equivalent pavement damages per pass relative to a standard 18-kip single axle (Morovatdar et al. 2019). Hence, EALF is the key component in the calculation of the ESAL values. Traditional industry-standard EALFs provided by the Asphalt Institute (AI), as well as the modified EALFs, were used to calculate the corresponding ESAL values, describing Level 2 traffic inputs.

Modified EALF values were determined using the mechanistic quantification of the pavement damages procedure described in Morovatdar et al. (2020a). Table 2 provides the modified EALF values attributed to the various roadway types, i.e., FM, SH, and US highways, for different axle weights and axle types of OW trucks operating in Texas overload corridors. As shown in Table 2, the modified EALFs that account for demanding loading conditions of OW vehicles, environmental conditions, and the unique features of transportation facilities in the network, were substantially higher than traditional EALFs currently employed by the pavement design industry. Additionally, the modified EALF values showed that in all axle types, FM roadways with less robust pavement profile had the highest damage factors among the three roadway types, followed by SH and, ultimately US highways. Detailed information regarding the approach, concepts, and the rationale for the calculations of the modified EALFs is provided in Morovatdar et al. (2020a).

Subsequent to the calculation of the damage equivalency factors, the number of axle load repetitions were derived from the site-specific axle load spectra databases. Ultimately, the EALF values for each axle type were multiplied by the projected frequencies of load repetitions to calculate the projected ESAL values. Figure 6 shows the cumulative ESAL values over 20-year design life for ten representative pavement sections in this study, based on modified and traditional damage equivalency factors. As expected, the modified ESALs were significantly higher than the traditional ESAL values calculated by using the AI damage equivalency factors.

Table 2. Modified and traditional industry-standard EALF values associated with OW Axles operating in Texas overload corridors

Axle Type	Axle Weight (kips)	Roadway Type			Traditional	Axle Type	Axle Weight (kips)	Roadway Type			Traditional
		FM	*SH*	*US*				*FM*	*SH*	*US*	
Single Axle	20	2.49	1.56	1.54	1.51	*Tandem Axle*	34	2.02	1.55	1.52	1.10
	22	3.49	2.29	2.22	2.18		36	2.47	1.92	1.87	1.38
	24	4.76	3.30	3.22	3.03		38	2.98	2.34	2.28	1.70
	26	6.34	4.63	4.44	4.09		40	3.57	2.82	2.67	2.08
	28	8.27	6.41	6.03	5.39		42	4.25	3.38	3.31	2.51
	30	10.62	8.66	8.17	6.97		44	5.00	4.00	3.80	3.00
	32	13.57	11.30	10.81	8.88		46	5.86	4.71	4.55	3.55
	34	17.05	14.80	14.05	11.18		48	6.81	5.50	5.29	4.17
	36	21.41	18.89	18.00	13.93		50	7.86	6.44	6.23	4.86
	38	26.85	23.86	22.76	17.20		52	9.05	7.42	7.25	5.63
	40	33.29	29.61	28.45	21.08		54	10.37	8.52	8.28	6.04

Axle Type	Axle Weight (kips)	Roadway Type			Traditional	Axle Type	Axle Weight (kips)	Roadway Type			Traditional
		FM	*SH*	*US*				*FM*	*SH*	*US*	
Tridem Axle	42	1.60	0.99	0.93	0.60	*Quad Axle*	50	1.42	0.86	0.82	0.60
	50	2.66	1.88	1.79	1.22		56	1.69	1.31	1.24	0.96
	58	4.48	3.26	3.11	2.20		62	2.26	1.91	1.83	1.45
	66	7.09	5.24	5.04	3.62		68	3.14	2.67	2.57	2.11
	74	10.64	7.97	7.71	5.57		74	4.23	3.65	3.53	2.96
	82	15.39	11.67	11.33	8.16		80	5.57	4.86	4.72	4.02
	90	21.51	16.45	16.08	11.55		86	7.14	6.38	6.19	5.31
	98	29.26	22.56	22.07	15.92		92	9.11	8.20	7.92	6.86
	106	38.90	30.11	29.57	21.48		98	11.38	10.33	9.99	8.70
	114	50.82	39.64	38.92	28.51		104	14.05	12.83	12.55	10.86
	120	76.23	59.47	58.38	42.77		110	17.22	15.81	15.49	13.39

3.3 Level 3 Traffic Data

The statewide average values, as Level 3 traffic input data, were used to incorporate the truck information in the comparative analysis. Figure 7 provides comparisons between Level 1 and Level 3 traffic data for quad-axle load distributions and truck class distributions at US 281 highway. Contrasting the two traffic data showed that the quad-axle load distributions in both case scenarios resembled a bell-shaped curve with a degree of skewness. However, TxME software tends to characterize extremely lower frequencies of OW loads compared to the actual field data collected at US 281 highway. Considering

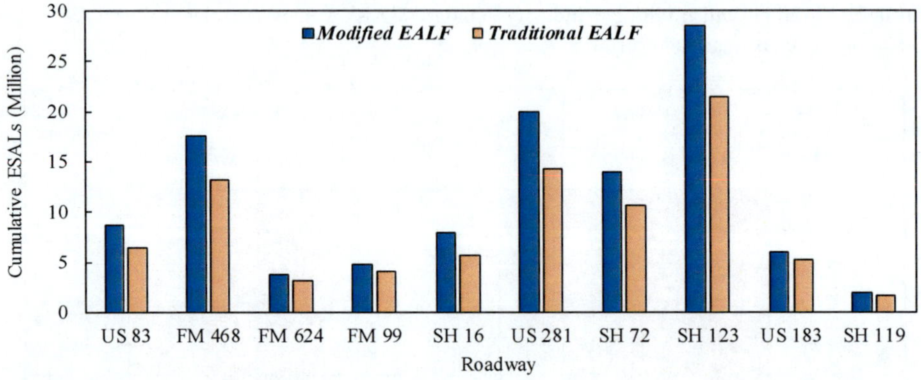

Fig. 6. Cumulative 18-kip ESAL values for representative roadways based on modified and traditional EALF values.

the Texas permissible weight limits for quad axles as 50 kips, OW axles with frequencies as high as 61% were characterized at US 281 based on real data recorded in the field; nevertheless, the corresponding percentage was determined to be 18% when Level 3 inputs are incorporated into the analysis.

Figure 7(b) also illustrates that the percentage of Class 9 trucks operating in US 281 was substantially higher than the corresponding percentage set forth by the software. It is noted that based on the research team's experience in relevant projects, Class 9 trucks often carry heavy loads and are characterized as OW and super-heavy trucks. For this reason, the frequency and operation of Class 9 trucks need to be accurately quantified, as they greatly contribute to the accumulated damages imparted on the pavements. Accordingly, the pre-defined traffic distribution patterns in ME software substantially deviate from the actual traffic mix and demanding loading conditions currently applied to the pavement facilities in such corridors with large volumes of OW truck operations.

Non-destructive Testing

GPR and FWD tests were deployed to the field, as shown in Fig. 8, for the determination of the layer configurations and layer moduli in representative pavement sections. This information was a direct input to the ME analysis protocol. Table 3 shows the pavement layer configurations after post-processing the collected data. Analyzing the GPR data validated by the pavement design plans showed that in the studied network, FM roadways, on average, consisted of only 1.6 in. (4.1 cm) of the asphalt treated surface layer. However, representative pavement sections of SH and especially US highways were found to be more robust, with an approximate average asphalt layer thicknesses of 4.8 in. (12.2 cm), and 6.3 in. (16 cm), respectively.

Additionally, the authors devised a plan to perform FWD testing in two different seasons (summer and winter), to properly evaluate the effect of environmental parameters on the back-calculated layer modulus of pavement layers. Based on the analysis of the FWD data for summer and winter seasons, the back-calculated layers modulus values in summer were substantially lower compared to wintertime modulus values from non-destructive testing of pavement sections. This is primarily attributed to the viscoelastic

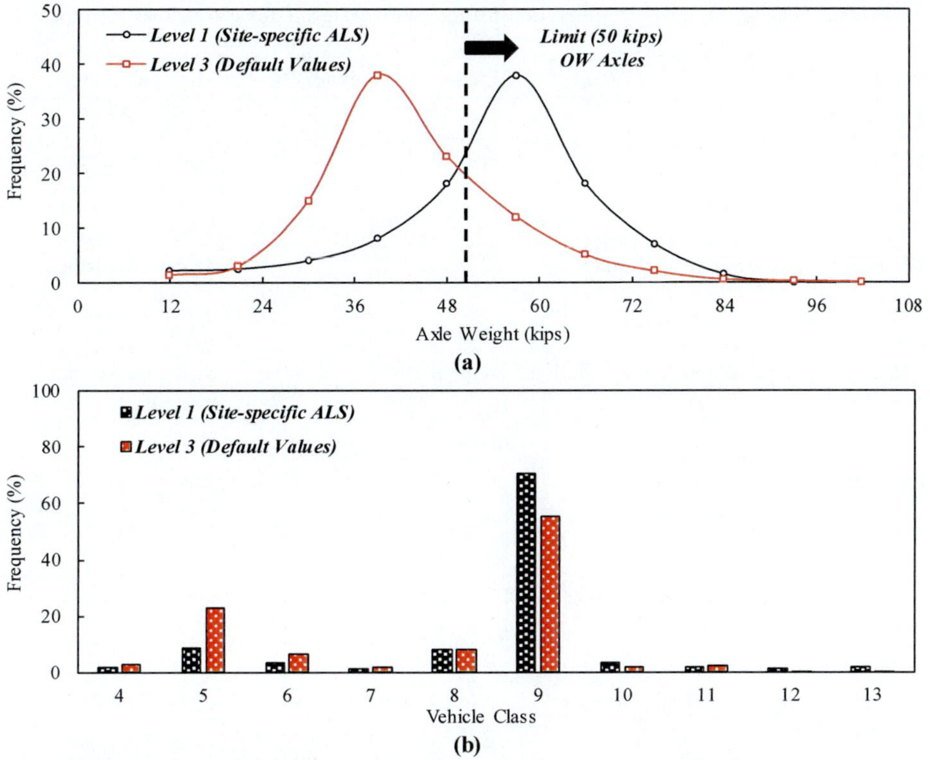

Fig. 7. Comparison of different traffic input scenarios at US 281 for (a) Quad axle load distributions, and (b) Truck class distributions.

Fig. 8. Non-destructive testing in Texas overload corridors (a) GPR testing, and (b) FWD testing.

nature of the asphalt layer, and softening of the surface layers due to elevated temperatures in summer seasons. More details on the back-calculated layer moduli associated with all ten representative sites are provided in Ashtiani et al. (2019).

Table 3. Pavement layers configurations of the studied sites in Texas overload corridors

Roadway	District	County	Layer	Thickness (in.)	Material
FM Road					
FM 468	Laredo	La Salle	Asphalt	2.1	Seal coat (surface treated)
			Base	9.8	UAB[a]
FM 624	San Antonio	McMullen	Asphalt	1.3	Seal coat (surface treated)
			Base	8.9	UAB
FM 99	San Antonio	McMullen	Asphalt	1.5	Seal coat (surface treated)
			Base	8.3	UAB
State Highway					
SH 16	San Antonio	Atascosa	Asphalt	3.5	HMA[b] (PG[c] 70-22)
			Base	7.4	UAB
SH 72	Corpus Christi	Karnes	Asphalt	1.5 + 5.1	Overlay + HMA (PG 64-22)
			Base	16	CTB[d]
SH 123	Corpus Christi	Karnes	Asphalt	5.5	HMA (PG 70-22)
			Base	15	UAB
SH 119	Yoakum	Dewitt	Asphalt	3.5	HMA (PG 64-22)
			Base	10.5	UAB
US Highway					
US 83	Laredo	Dimmit	Asphalt	1.5 + 4.8	Overlay + HMA (PG 70-22)
			Base	6.2	UAB
US 281	Corpus Christi	Live Oak	Asphalt	1.5 + 5.5	Overlay + HMA (PG 70-28)
			Base	18	UAB
US 183	Yoakum	Gonzales	Asphalt	5.5	HMA (PG 70-22)
			Base	8	UAB

[a] Unbound Aggregate Base
[b] Hot Mix Asphalt
[c] Performance Grading
[d] Cement Treated Base

4 Field Data Analysis and Results

4.1 Influence of Traffic Characterization Methodology on Service Life Predictions

Figure 9 shows the progression of rutting in the heavily trafficked US281 highway in Corpus Christi District, based on different traffic input scenarios evaluated in this study. Analysis of the various types of pavement distresses in US 281 indicated that the rutting was the critical pavement distress controlling the pavement life at this site. For this reason, the rutting performance of US 281 based on various traffic inputs was studied for further comparison purposes. It is also noted that since the evaluated section in US 281 was reconstructed in 2014, the onset of the analysis (baseline) was assumed in that year.

As evidenced in Fig. 9, analysis of the rutting performance over a 20-year design period for US 281 indicated that if the site-specific ALS data (Level 1), was incorporated into the ME analysis procedure, it takes 133 months for this roadway to reach failure criterion of 0.5 in. However, if the calculated ESAL values, based on the modified and traditional industry-standard damage equivalency factors, were incorporated into the analysis, it takes 150 and 187 months, respectively, to develop 0.5 in. of cumulative rut depth. In other terms, simulations with site-specific ALS resulted in significantly lower predictions of service life when compared to those determined when using the traditional ESAL concept. This is in line with our expectations, as the damage quantification based on the ESAL concept overlooks the load groups and time-dependency of the traffic distributions. For this reason, incorporation of ESALs into the ME analysis can underestimate the imparted pavement damages by OW truck operations, leading to an overestimation of the pavement service life in the overload corridors.

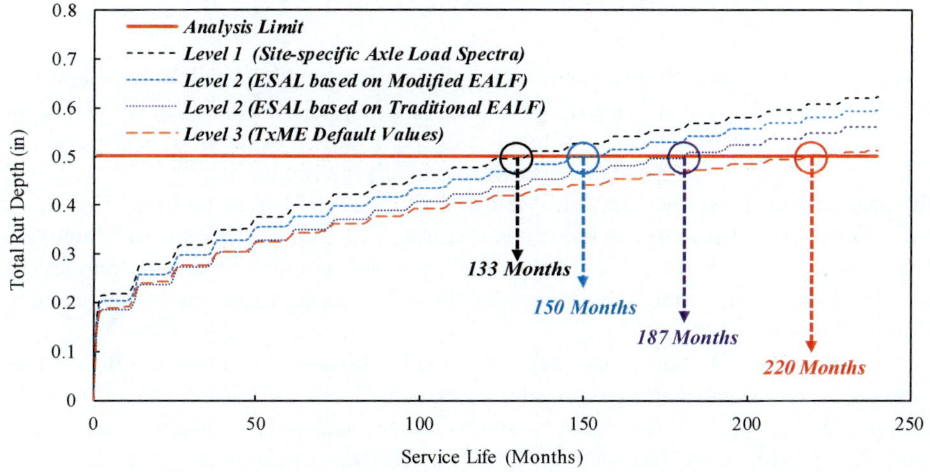

Fig. 9. Predicted service life at US 281 highway using different traffic inputs.

Another method to assign traffic characteristics is by using the mechanistic software average values, as commonly used for pavement management purposes in practice. The post-processed results illustrated in Fig. 9 show that using the default traffic distributions as Level 3 traffic inputs resulted in substantial deviations from the traditional ESAL concept and site-specific ALS databases. This underscores the significance of using site-specific traffic data in lieu of software default values for decision-making purposes.

4.2 Comparisons of Predicted and Field-Measured Rut Depth

The predicted rutting values from the proposed mechanistic approach were contrasted with those obtained from field distress surveys of the representative pavement sections to assess the accuracy of the analysis results based on the site-specific ALS. Rutting performance of US 281 is discussed in this section as a case study. As shown in Fig. 9, our methodology predicted a rut depth of 0.380 in. (9.65 mm) on US 281 by spring 2019 after five years of service. This is in line with the rut depth measurements conducted along US 281, as shown in Fig. 10. The pictures provided in Fig. 10 show rut depth measurement under the straight edge in the wheel path to be approximately 0.375 in. (9.53 mm) on average, in the spring of 2019.

Fig. 10. Rutting measurements in US 281, Corpus Christi district.

Figure 11 illustrates the measured versus predicted rut depths for all sites in this study. The Root Mean Square of Error (RMSE) between the predicted and field-measured rut depths were calculated for cross-validation of the accuracy of the proposed algorithm for service life analysis. The results show that incorporating the site-specific ALS into the analysis for the studied sections yielded a rut depth RMSE of 65 mils (1.65 mm). Such low measures for error in pavement distress predictions, coupled with a relatively high R-squared value of 0.90 of the fitted line, provides confidence in the soundness of the algorithm developed for the assessment of the service life of pavements subjected to OW trucks in this study.

Additionally, a student t-test was performed to examine whether the differences between predicted and measured rut values were statistically significant. The null hypothesis was defined if the differences in predicted and measured rut values were equal to zero. A significance level (α) of 0.05 for 95% confidence was assumed, implying that a p-value of 0.05 or less rejects the null hypothesis. Based on the student's one-sided t-test, the p-value was found to be 0.66 (>0.05). Therefore, the null hypothesis was accepted, indicating that there is no significant arithmetic difference between the predicted and measured rut depth values.

Comparisons between the predicted and field-measured rutting indicated that the proposed methodology for prediction of the service life, which is based on the site-specific ALS, is capable of assessing the incremental progression of the distresses imparted during the service life of the pavements.

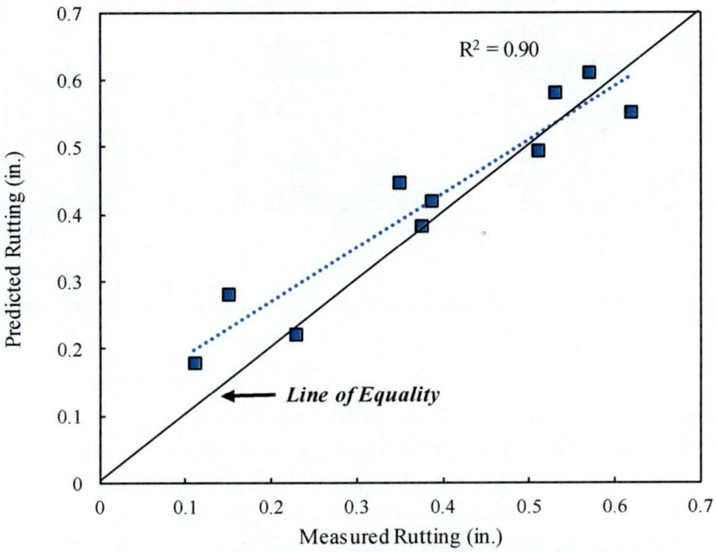

Fig. 11. Predicted vs. field-measured surface rutting.

4.3 Overestimation of Pavement Service Life due to Level 2 and Level 3 Traffic Inputs

In order to clarify the influence of traffic inputs on the pavement service life predictions, the results are reported as "percent overestimation," indicating the corresponding overestimation of service life when Level 2 and Level 3 data, instead of Level 1 traffic data, is incorporated into the analysis. Figure 12 shows the comparative results for US281, SH123, and FM468 roadways based on different traffic characterization methodologies. As evidenced in the plot, in all three roadway types, incorporation of both ESAL and software default values into the ME analysis resulted in an overestimation of the service life. This is due to the significant deviations of the actual traffic loading conditions in overload corridors from the corresponding loading conditions in Level 2 and Level 3 traffic data. Consequently, the procedures that merely rely on Level 2 and Level 3 traffic inputs can potentially jeopardize the accuracy of the analysis of the pavement service life in the overload corridors.

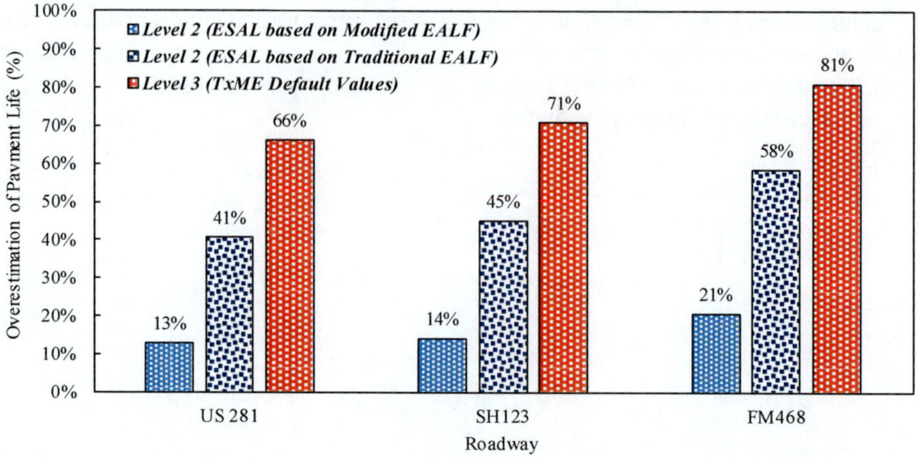

Fig. 12. Overestimation of service life for different pavement types based on different traffic inputs.

It was also observed that the default traffic inputs lead to the riskiest pavement life predictions. Similarly, the traditional ESAL values resulted in 44% to 58% overestimation in the service life of the pavements. However, incorporation of the ESAL values calculated based on the modified damage equivalency factors into the analysis resulted in the closest service life predictions to those predicted using site-specific ALS. The relative differences between the corresponding results ranged from 13% to 21%. Therefore, using the modified damage equivalency factors for calculation of the ESAL values in the analysis can provide an initial estimate of the pavement service life, when the field-derived ALS is unavailable.

Another noteworthy observation from Fig. 12 pertains to the relevance of the type of transportation facility and the overestimation of the pavement service life. In the studied sections, such overestimation was more pronounced for FM 468, compared to SH 123 and US 281 highways. This is primarily attributed to the fact that the deviation of the field-characterized traffic patterns from the software pre-defined traffic distributions was more substantial at FM roadways. Accordingly, using the traditional ESAL concept and software default values to characterize traffic in ME analysis resulted in the riskiest pavement life predictions for FM roadways, compared to other studied pavement types.

5 Summary and Conclusions

In this study, the influence of different traffic characterization methodologies on the performance and service life of the pavements subjected to the OW truck operations was accurately assessed. Four traffic input scenarios, including site-specific ALS (level 1), ESAL based on traditional EALF (level 2), ESAL based on modified EALF (level 2), and ultimately software default values (level 3) were considered in the comparative analysis conducted in this study. Ten representative pavement sections consisted of FM, SH, and US highways were selected for field traffic monitoring using P-WIM devices. The traffic data collected by P-WIM units were then synthesized to develop the site-specific ALS databases. Subsequently, using the damage equivalency factors, the characterized traffic mix was converted to the equivalent 18-kip axles based on the traditional ESAL concept. A series of non-destructive field tests, including GPR and FWD at two intervals, were incorporated in this study. Ultimately, different traffic input scenarios, structural properties, and climate information were in turn incorporated into a series of numerical simulations in TxME for the determination of the service life of the pavements subjected to taxing traffic loading conditions. The major findings of this research effort summarized as:

- An accurate account of traffic characteristics is the precursor for reliable assessment of the service life of pavements in overload corridors. Site-specific ALS data provides the primary mechanistic traffic data input for accurate and optimal pavement design and analysis.
- Comparative analysis of different traffic characterization methodologies, i.e., ALS, and traditional ESAL concept, underscored the significance of using the ALS database in lieu of traditional conversion of traffic mix to standard axle for the prediction of the service life of pavements.
- Incorporation of ESALs instead of ALS, in service life analysis, vastly overestimated the service life of the pavements. This is primarily attributed to the fact that the damage quantification based on the ESAL concept overlooks the load groups and time-dependency of the traffic distributions, leading to an underestimation of the imparted pavement damages by OW truck operations.
- Using the modified mechanistic-based damage equivalency factors that account for the specific characteristics of OW vehicles can provide the closest service life predictions to those determined using the site-specific ALS.
- Using the average statewide traffic distributions to characterize the traffic mix resulted in drastic overestimation of the predicted service life compared to traditional ESAL concept and site-specific ALS databases. This highlights the importance of using site-specific traffic data rather than using the software/agency default values for decision-making purposes.

References

Abbas, A.R., Frankhouser, A., Papagiannakis, A.T.: Effect of traffic load input level on mechanistic-empirical pavement design. Transp. Res. Rec. J. Transp. Res. Board **2443**(1), 63–77 (2014)

Al-Qadi, I., et al.: Development of a Proposed Overweight Vehicle Permit Fee Structure in Illinois. Publication FHWA-ICT-17–004. Illinois Center for Transportation (2017)

Ashtiani, R.S., Morovatdar, A., Licon, C., Tirado, C., Gonzales, J., Rocha, S.: Characterization and Quantification of Traffic Load Spectra in Texas Overweight Corridors and Energy Sector Zones. No. FHWA/TX-19/0-6965-1 (2019)

Banerjee, A., Prozzi, J.A.: Practical approach for determining permit fees for overweight trucks. Transp. Res. Rec. **2478**(1), 93–102 (2015)

Batioja-Alvarez, D.D., Kazemi, S.F., Hajj, E.Y., Siddharthan, R.V., Hand, A.J.: Probabilistic mechanistic-based pavement damage costs for multitrip overweight vehicles. J. Transp. Eng. Part B Pavements **144**(2), 04018004 (2018)

Chou, E., Pulugurta, H., Datta, D.: Pavement Forecasting Models. Publication FHWA/OH-2008/3. Ohio Department of Transportation (2008)

Haider, S.W., Buch, N., Chatti, K., Brown, J.: Development of traffic inputs for mechanistic-empirical pavement design guide in Michigan. Transp. Res. Rec. J. Transp. Res. Board **2256**(1), 179–190 (2011)

Jasim, A.F., Wang, H., Bennert, T.: Evaluation of clustered traffic inputs for mechanistic-empirical pavement design: case study in New Jersey. Transp. Res. Rec. J. Transp. Res. Board **2673**(11), 332–348 (2019)

Li, Q., Wang, K.P., Eacker, M., Zhang, Z.: Clustering methods for truck traffic characterization in pavement ME design. ASCE-ASME J. Risk Uncertainty Eng. Syst. Part A Civ. Eng. **3**(2), F4016003 (2016)

Mai, D., Turochy, R.E., Timm, D.H.: Correlation-based clustering of traffic data for the mechanistic-empirical pavement design guide. Transp. Res. Rec. J. Transp. Res. Board **2339**(1), 104–111 (2013)

Morovatdar, A., Ashtiani, R.S., Licon, C., Tirado, C., Mahmoud, E.: Novel framework for the quantification of pavement damages in the overload corridors. Transp. Res. Rec. (TRR) J. Transp. Res. Board. **2674**, 179 (2020a). https://doi.org/10.1177/0361198120925807

Morovatdar, A., Ashtiani, R.S., Licon, C.: Development of a mechanistic framework to predict pavement service life using axle load spectra from Texas overload corridors. In: ASCE's International Conference on Transportation & Development, Seattle, Washington (2020b)

Morovatdar, A., Ashtiani, R.S., Licon, C., Tirado, C.: Development of a mechanistic approach to quantify pavement damage using axle load spectra from South Texas overload corridors. In: Geo-Structural Aspects of Pavements, Railways, and Airfields Conference (GAP 2019), Colorado Springs, CO, USA (2019)

Oh, J., Walubita, L.F., Leidy, J.: Establishment of statewide axle load spectra data using cluster analysis. KSCE J. Civ. Eng. **19**(7), 2083–2090 (2015)

Romero, P., Bao, S., Sudbury, D.: Performance Evaluation of Typical UDOT Surface Treatments. No. UT-19.17 (2019)

Rys, D., Judycki, J., Jaskula, P.: Analysis of effect of overloaded vehicles on fatigue life of flexible pavements based on weigh in motion (WIM) data. Int. J. Pavement Eng. **17**(8), 716–726 (2016)

Smith, B.C., Diefenderfer, B.K.: Analysis of virginia-specific traffic data for use with mechanistic-empirical pavement design guide. Transp. Res. Rec. J. Transp. Res. Board **2145**, 100–107 (2010)

Walubita, L.F., Lee, S.I., Djebou, C., Aldo, A., Prakoso, A.: Using WIM Systems and Tube Counters to Collect and Generate ME Traffic Data for Pavement Design and Analysis. Publication FHWA/TX-18/0-6940-R1. FHWA, U.S. Department of Transportation (2019)

Wang, H., Zhao, J., Wang, Z.: Impact of overweight traffic on pavement life using weigh-in-motion data and mechanistic-empirical pavement analysis. In: 9th International Conference on Managing Pavement Assets (2015)

Determination of Landslide High Risk Areas Using GA and GIS Combination in the West of Mazandaran Province

Reza Aghababaee Pour, Hossein Etemadfard(✉), and Rouzbeh Shad

Civil Engineering Department, Faculty of Engineering, Ferdowsi University of Mashhad (FUM), Mashhad, Iran
{etemadfard,r.shad}@um.ac.ir

Abstract. The natural hazards have an important role in the human life and their damages may compensate more than several years. Landslides, as the one of the natural disasters, have killed over 107 people and damaged about 386 billion Rials in Iran that happened from 2548 landslides. Based on this impact, this study intends to determine the landslide high-risk areas in the west of Mazandaran. For this purpose, the required and available layers have been determines which are the parameters of slope, land use, distance from the fault, distance from the river, distance from the road, rainfall, and geology. According to the estimated effects and damages, each layer was classified into the risk levels. The Genetic Algorithm was implemented on the layers in order to determine the dangerous areas. The input data were divided to non-fuzzy and fuzzy sets to detect the high-risk areas in two methods. Then, the high-risk susceptible sites were specified on the map for both of them. The obtained results were assessed and compared non-fuzzy and fuzzy approaches. This shows that the northern part of the region has the landslide high-risk areas rather than other regions in the west of Mazandaran province.

Keywords: Natural hazard · Evolutionary algorithms · Landslide · Geoscience

1 Introduction

Natural hazards and risks have become an important issue affecting society safety and their damage to human life and the environment cannot be ignored (Chen et al. 2018a). Landslides are a major natural hazard that often damages to properties and lives, and led to the high reconstruction costs. (Chen et al. 2016). Landslide susceptibility can be defined as the spatial probability of the landslide occurrences according to a set of geological and environmental conditions (Chen et al. 2017). Landslide susceptibility mapping requires a multi-criteria approach and high levels of accuracy and reliability in the resulting maps, in order to be relevant for decision-making and the design of disaster management plans (Feizizadeh et al. 2014). Because of the different aspects of landslides,

A. Tapase et al. (Eds.): GeoChina 2021, SUCI, pp. 75–88, 2021.
https://doi.org/10.1007/978-3-030-79644-0_7

such as economic, social, environmental, and health, the landslide hazard zoning maps are so important for governmental and political sections to identify sensitive areas and sustainable locations in the future development planning (Chen et al. 2018a). These are shown the necessity of landslide susceptibility assessment to facilitate forecasting of this phenomenon (Chen et al. 2018b).

Iran is one of the countries that generally faces landslides and suffers a lot of losses every year. In Iran, studies have been conducted on the damage caused by landslides, the results of which show that 4900 landslides have been recorded in Iran by the end of September 2007, causing damage to 126893 billion Rials (Ebrahimi et al. 2015). Mazandaran province with a special climate due to the confinement between the Caspian Sea and the Alborz Mountains has the high landslide occurrences. On the other hand, landslide factors such as proper rainfall along with erosion-sensitive formations, as well as being on the Alpine-Himalayan mountain belt and active faults, have provided the necessary potential for landslides in this province (Pourghasemi et al. 2016).

In recent decades, with the development of technology for measuring, collecting observations and knowledge of data analysis, many studies have been conducted in the fields of quantitative and qualitative investigation of landslides (Chen et al. 2017). Based on the development, researchers who work in the landslide susceptibility mapping fields have applied GIS (Geospatial Information System) to assess and spatial analysis by using algorithms such as logistic regression (Nandi et al. 2009; Yalcin et al. 2011; Bai et al. 2010; Chen et al. 2016; Aditian et al. 2018; Ozdemir et al. 2012), kernel logistic regression (Chen et al. 2018a; Hong et al. 2015), Artificial Neural Network (ANN) (Bragagnolo et al. 2019; Chen et al. 2016; Thai Pham et al. 2015; Thai Pham et al. 2017; Aditian et al. 2018), frequency ratio (Khan et al. 2018; Yalcin et al. 2011; Aditian et al. 2018; Ozdemir et al. 2012; Vakhshoori and Zare 2016; Jaafari et al. 2013), Analytic Hierarchy Process (AHP) (Feizizadeh et al. 2012; Yalcin et al. 2011; Shahabi et al. 2015), weights of evidence (Ozdemir et al. 2012; Hong et al. 2017; Saha et al. 2020; Poonam et al. 2016), Support Vector Machine (SVM) (Saha et al. 2020; Hong et al. 2015; Peng et al. 2013).

As can be seen in the above research, the zoning of high-risk landslide areas has been analyzed using many data and algorithms, but the use of evolutionary algorithms has not received much attention. In this study, Genetic Algorithm (GA), as an evolutionary algorithm, and its combination with fuzzy logic have been applied for that purpose. As pointed above, the risk of landslides in northern Iran is high, so the area was defined as a study area. In the other word, the main purpose of this study is to investigate landslide high risk areas using GA with two types of fuzzy and non-fuzzy data sets.

2 Methodology

2.1 GA

GA is inspired by Darwin's theory of evolution (Saeidian et al. 2016) and the hypothetical mechanism of natural selection where the fittest individuals at one generation are more likely to survive and produce the new generation (Kavzoglu et al. 2015). The algorithm has already been widely used for different optimization and machine learning applications (Saeidian et al. 2016). GA is used to search the optimal solutions when the evaluation of all possible solutions is too costly in terms of computing time (Kavzoglu et al. 2015). The building blocks of GA are the genes or solution parameters. Each solution to the problem is called a chromosome and is composed of some genes (Saeidian et al. 2016).

One of the main stages of GA is the calculation of the fitness (objective) functions to find the solutions. This is the only criteria for selection of the best solutions that are to bc parents of the next generation. The solutions are selected to be parents with probabilities proportionate to their fitness values. The more often used methods of selection are the roulette wheel, tournament and ranking methods. In the crossover operation, sections of two parents are switched and combined to generate two new solutions. Such new solutions might have better fitness values than their parents. Many methods for the crossover are available, including Ranking Selection, N-Point, Uniform, Partially Mapped, Ordered, Cycle, and Diagonal Crossover. In order to prevent the convergence of algorithm in local optimum points, and to search new areas of the search space, the mutation operation is used. It makes small changes in some genes of a few chromosomes (Saeidian et al. 2016). The basic stages of GA are depicted in (Fig. 1).

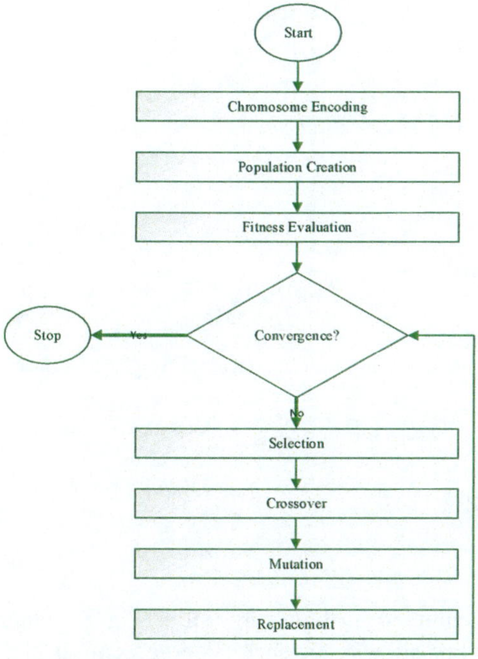

Fig. 1. Flowchart for GA (Saeidian et al. 2016).

2.2 Fuzzy Logic Model

In classical set theory, the membership value of an element in a set is considered to be 0 or 1 (each element either belongs to the set or does not belong to it). According to fuzzy set theory, a fuzzy set is a subset whose membership value of the main set is between zero and one with respect to an intermediate membership function (Zadeh 1965). The membership functions are different, the linear membership function has been applied here. The linear membership function μ_L (Kumar Singh and Prasad Yadav 2017) can be defined as follows;

$$\mu_L(Z_p(x)) = \begin{cases} 1, & \text{if } Z_p \leq L_p, \\ \frac{U_p - Z_p}{U_p - L_p}, & \text{if } L_p \leq Z_p < U_p, \\ 0, & \text{if } Z_p \geq U_p, \end{cases} \tag{1}$$

This function is a strictly decreasing concave and convex function for $Z_p(x)$. Figure 2 shows the linear membership function.

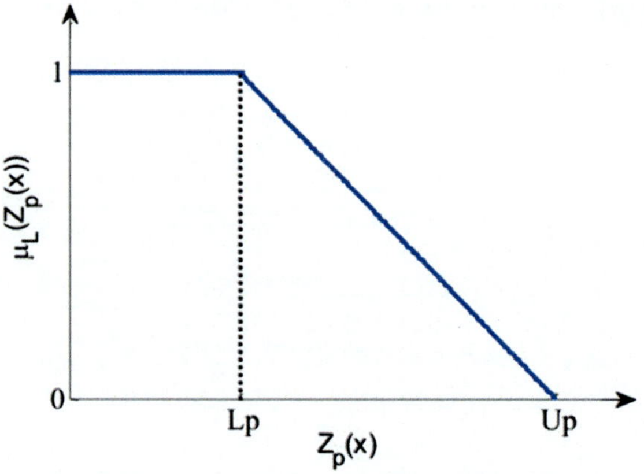

Fig. 2. Linear membership function (Kumar Singh and Prasad Yadav 2017).

3 Area and Data

3.1 Study Area

The study area is located in the northern part of Iran (Fig. 3), which is one of the highest risk areas in Iran for landslide occurrence (Pourghasemi et al. 2012), because of its climate, geology, hydrogeological environment, geomorphological conditions and high tectonic activities. Human activities such as deforestation , timber harvesting in slope imbalance and road construction in forest lands are other causes of landslides in the study region (Arab Amiri and Conoscenti 2017). The area lies between 36° 09′ 14″ N and 36° 58′ 36″ N latitude and 50° 20′ 16″ E and 51° 58′ 02″ E longitude (the west of Mazandaran), and covers an area about of 6100 km^2 (Fig. 3). The altitude varies from −30 to 4741 m and increases from the north to the south. A number of landslides have occurred in the past; there are also numerous recorded instances of active landslides in this area.

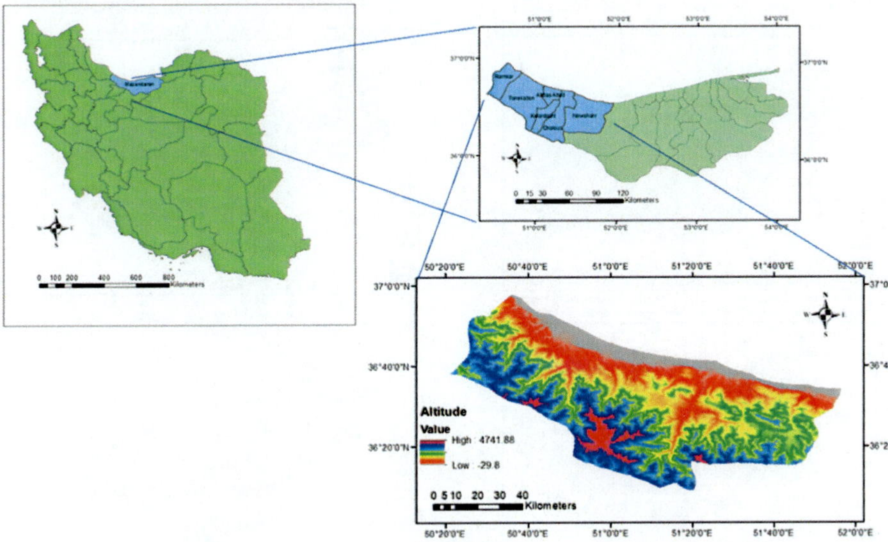

Fig. 3. Location of the study area in the west of Mazandaran

3.2 Input Data

To detect the high-risk areas, criteria related to landslide occurrence should be determined and collected. After reviewing the previous studies and gathering the available data, the geological layers, land use, slope, distance from the fault, rainfall, distance from the river and distance from the road have been prepared in shape file format (shp) using ArcGIS. Each parameter has been classified and the score of class indicates the risk of landslides. The distance to fault, geology classification, rain contours, land-use parcels, distance to river, distance to road; and slope degree classification have been illustrated in Fig. 4, respectively.

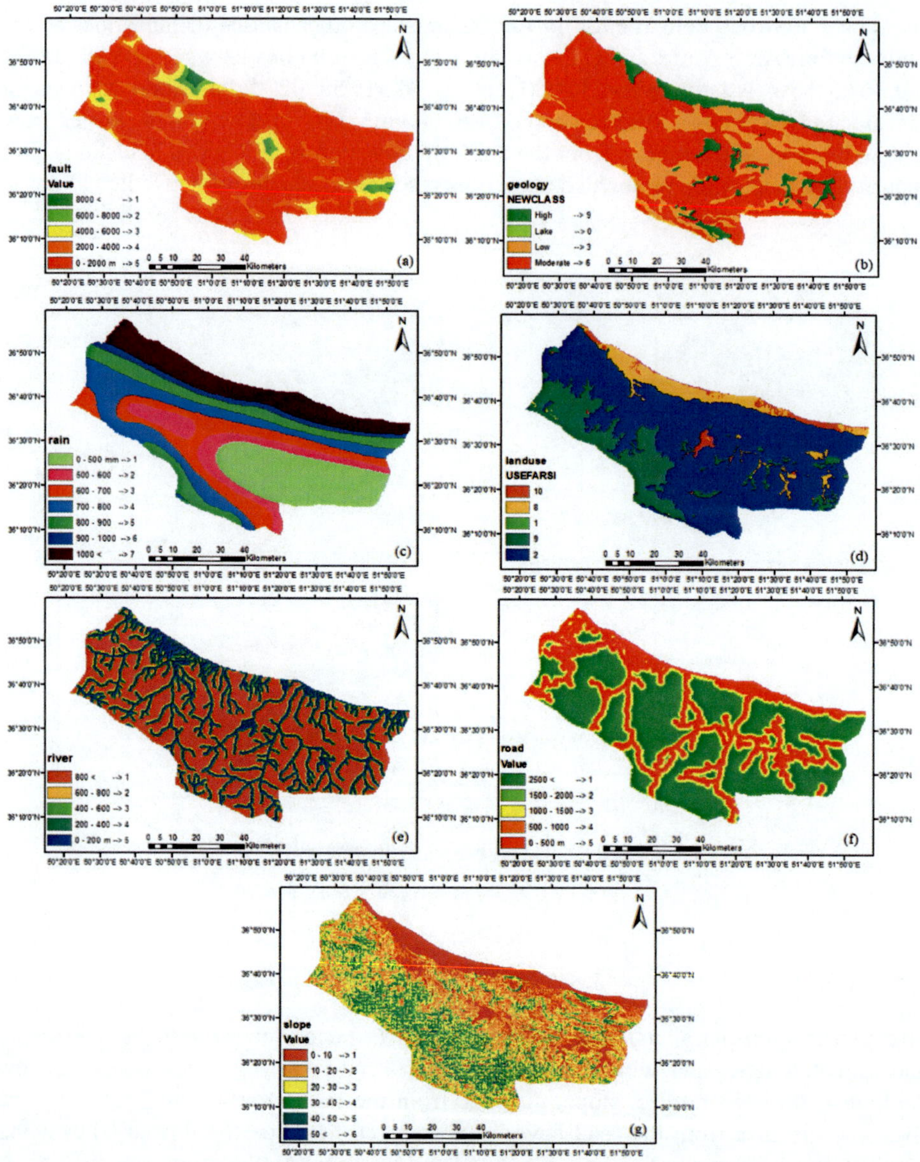

Fig. 4. Thematic maps of the study area: (a) distance to fault; (b) geology; (c) rain; (d) landuse; (e) distance to river; (f) distance to road; (g) slope degree.

As mentioned above, two sets of fuzzy and non-fuzzy data have been used to implement the GA in the study area. The data, that has been shown in Fig. 4, is converted to the fuzzy data set using fuzzy logic and the interval of the score of each layer will be changed between zero and one. For this purpose, fuzzy membership analysis in ArcGIS software and linear membership function have been applied to determine membership of each area and layers' objects. The fuzzy data shows in Fig. 5.

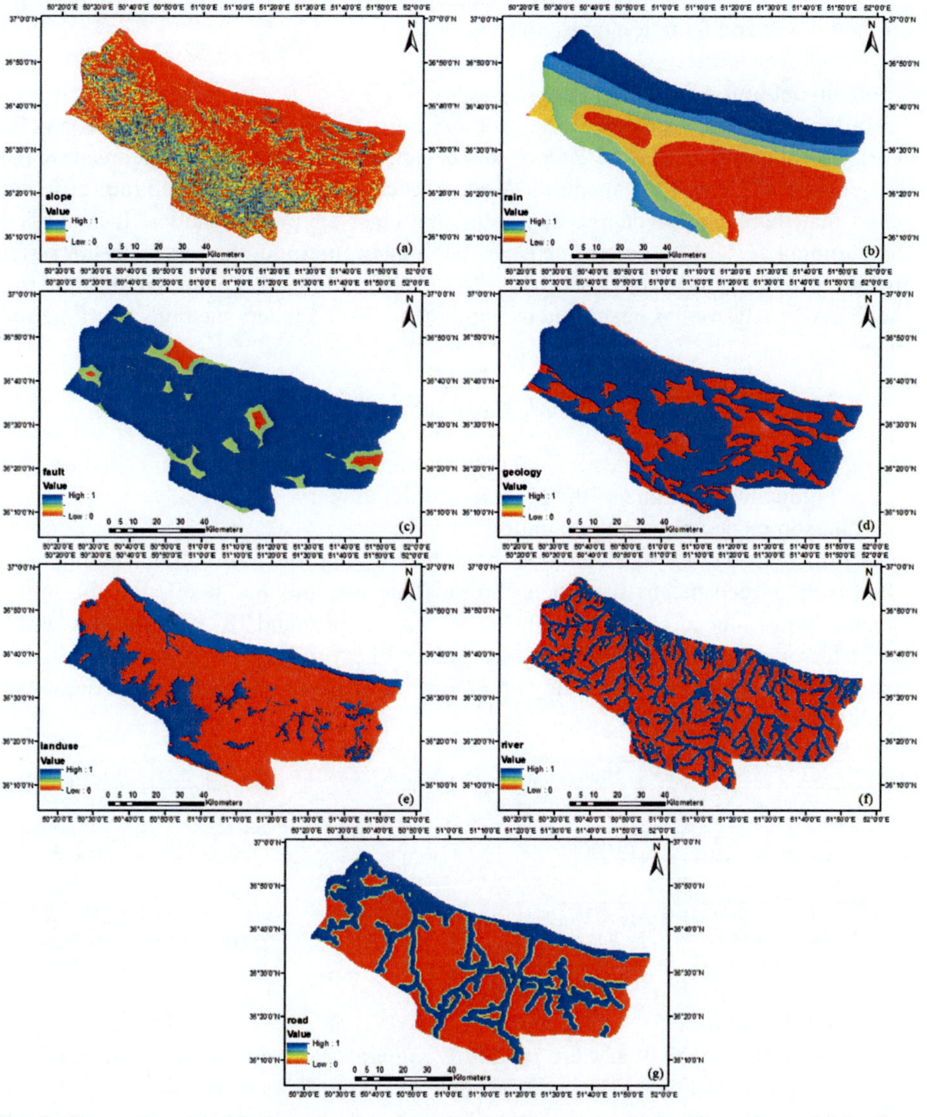

Fig. 5. Fuzzy data set: (a) slope degree; (b) rain; (c) distance to fault; (d) geology; (e) landuse; (f) distance to river; (g) distance to road.

4 Implementation

This section presents the steps for implementing the research. The problem is defined first and then the initial population is created. After that, the fitness function evaluates chromosomes and the best ones will be selected. The new generation is produced by using crossover and mutation operation.

- Problem definition
  In order to implement the GA, the problem must first be properly identified and its cost function must be created. Zoning and detection of landslide high risk areas are the target of this research. As mentioned above, each factor was classified into different class that indicates the degree of risk of that class against landslides. It is applied as the input for GA to determine high risk areas. The input data were in the raster format with the same number of pixels (1873 * 2873) and equal pixel size (50 * 50 m). MATLAB software has been used to implement the GA where the objective function was:

$$Fitness(i,\ j) = SUM\,(a_1,\ a_2,\ a_3,\ a_4,\ a_5,\ a_6,\ a_7) \tag{2}$$

  In Eq. 2, (i, j) represents the row and column number of the desired pixel and $(a_1, a_2, ..., a_7)$ indicates the value of the pixel in different layers.
- Chromosome encoding and population creation
  The primary chromosome contains the values of the row and column numbers of the pixels, which means that each chromosome contains one pixel, and the initial population produced on this chromosome is between 0 and 1873 for the row and 0 to 2873 for the column. If these two are placed next to each other, they represent a pixel of 50 × 50 m from the study area. For the highest level each chromosome has 550,000 genes (pixels).

586	761	320	1613	1028	....	1802	124	981	846

2621	202	1957	576	2029	....	1413	915	776	2361

586 2621	761 202	320 1957	1613 576	1028 2029	....	1802 1413	124 915	981 776	846 2361
1	2	3	4	5	1047000	1048000	1049000	1050000	

- Fitness evaluation
  In this step, the pixels that were randomly generated on the chromosome enter the fitness function and the risk of each pixel against landslides is calculated.

25	36	13	41	19	....	32	39	40	16

- Selection
  Here, the output values of the fitness function become a column matrix. The values of this matrix are arranged from high to low, with higher values indicating higher

landslide risk. Out of 1,050,000 pixels, the 350,000 pixels with the highest value go straight to the next iteration, the next 420,000 pixels go into the crossover and mutation phase, and the rest of the pixels are removed and replaced with new non-repetitive pixels.

- Crossover and mutation
In this step, assuming that pixels close together can behave similarly (a pixel from a high-risk area, its neighbors can be similar), the last row and column number of the two pixels are moved together and new pixels are created.

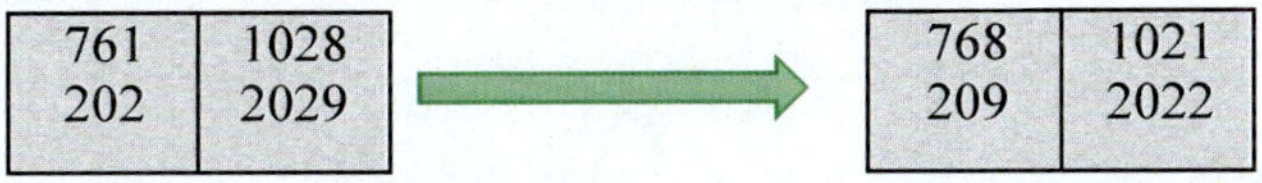

- Completion condition
Finally, if the program can't get better results after 100 iterations (higher risk pixels), the loop is stopped and the last answer is considered as the final answer.

5 Numerical Result

After implementing GA with the input data, two sets of output have been calculated which are non-fuzzy and fuzzy. The output is the landslide risk number (equal to the fitness function) that calculated for each pixel. Pixels with higher values belong to landslide high-risk areas. Four classes have been considered selecting the most high-risk areas. The Table 1 shows the number of pixels, areas and their percentage in those classes.

Table 1. High-risk zone area in four classes

Number of pixels	Area (hectare)	Percentage of study area
50,000	12,500	2.04%
100,000	25,000	4.09%
150,000	37,500	6.13%
250,000	62,500	10.22%

The spatial distribution of high-risk pixels are important to study the main landslide parameters and determine the high-risk areas for prevention operations. For this purpose, the distributed maps have been prepared which show the high-risk regions in red color. Figure 6 illustrated the high-risk maps in the four classes and non-fuzzy and fuzzy solution.

The maps indicate that the northern part of the case study has the most landslide high-risk where is close to the sea. High rainfall, closer proximity to roads and rivers, urban land use, garden and irrigated land and most importantly the geological type of the region have led to this conclusion. With the expansion of high-risk areas or increase the numerous high-risk pixels, there are areas in other parts of the region. It can be seen that the landslide high-risk areas have emerged in the south of the region because of slope.

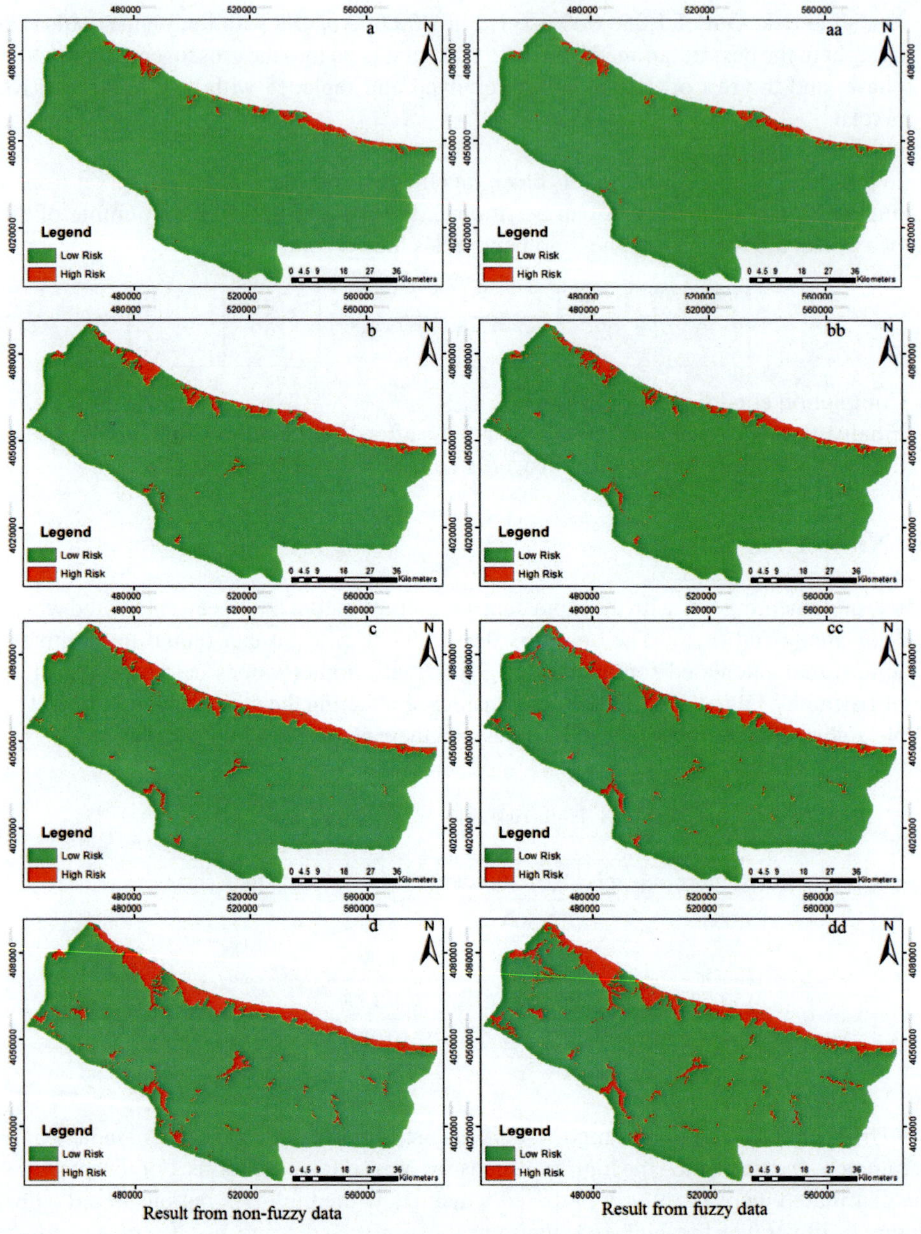

Fig. 6. Landslide High Risk Areas: a) 50,000 pixels non-fuzzy data; aa) 50,000 pixels fuzzy data; b) 100,000 pixels non-fuzzy data; bb) 100,000 pixels fuzzy data; c) 150,000 pixels non-fuzzy data; cc) 150,000 pixels fuzzy data; d) 250,000 pixels non-fuzzy data; dd) 250,000 pixels fuzzy data.

In addition, the results of fuzzy and non-fuzzy data are different in some areas. Based on Fig. 6, the high-risk areas in the northern strip of the region are more than fuzzy data, but the fuzzy data have identified more areas in other parts of the case study. As it can be seen in the maps, the high-risk areas in the western region are in the results of fuzzy data, it means that they are less on the results of non-fuzzy data. Figure 7 were prepared to enhance the differences between fuzzy and non-fuzzy data.

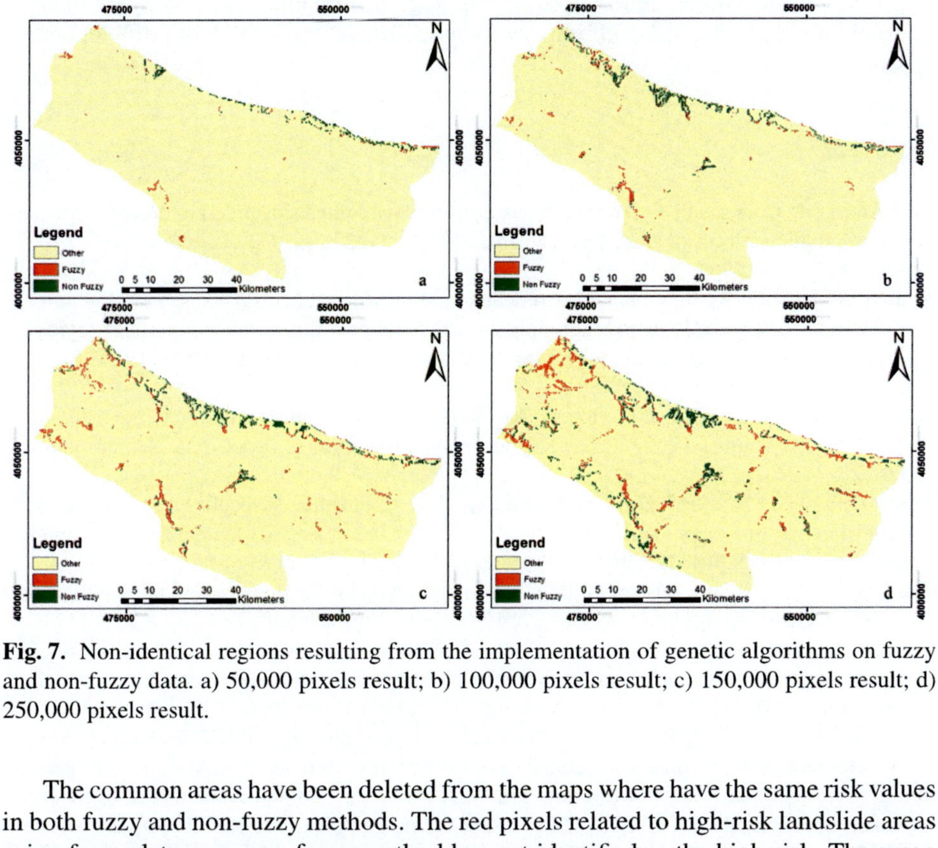

Fig. 7. Non-identical regions resulting from the implementation of genetic algorithms on fuzzy and non-fuzzy data. a) 50,000 pixels result; b) 100,000 pixels result; c) 150,000 pixels result; d) 250,000 pixels result.

The common areas have been deleted from the maps where have the same risk values in both fuzzy and non-fuzzy methods. The red pixels related to high-risk landslide areas using fuzzy data were non-fuzzy method has not identified as the high-risk. The green pixels are in the vise-versa condition. According to the maps, the fuzzy method enhanced the high-risk areas around roads and rivers. In the other word, it increases the weight of parameter that related to roads and rivers.

6 Conclusion

In this research, GA, as the one of the optimization algorithms, has been used to determine the landslide high-risk areas in GIS environment in the western part of Mazandaran province. The algorithm applied to the input data in non-fuzzy and fuzzy types. It's implemented in MATLAB software and the results were presented in two groups and in a number of different selected pixels.

The results show that there are areas with more landslide risk in the northern regions of the region and smaller areas were distributed throughout the region. It is noteworthy that there are differences between the results of fuzzy data and non-fuzzy data. These changes are more pronounced in the southern part of the region, which has more roads and rivers.

It is suggested that in future studies, other optimization algorithms be used to determine areas with high landslide risk and also for other natural hazards. Comparing different algorithms can be very helpful in this regard. Reliable results can be obtained by combining optimization algorithms with modeling methods such as artificial neural networks.

References

Arab Amiri, M., Conoscenti, C.: Landslide susceptibility mapping using precipitation data, Mazandaran province, North of Iran. Nat. Hazards **89**(1), 255–273 (2017). https://doi.org/10.1007/s11069-017-2962-8

Aditian, A., Kubota, T., Shinohara, Y.: Comparison of GIS-based landslide susceptibility models using frequency ratio, logistic regression, and artificial neural network in a tertiary region of Ambon, Indonesia. Geomorphology **318**(2018), 101–111 (2018). https://doi.org/10.1016/j.geomorph.2018.06.006

Bai, S., Wang, J., Lu, G., Zhou, P., Hou, S., Xu, S.: GIS-based logistic regression for landslide susceptibility mapping of the Zhongxian segment in the Three Gorges area, China. Geomorphology **115**, 23–31 (2010). https://doi.org/10.1016/j.geomorph.2009.09.025

Bragagnolo, L., Silva, R.V.D., Grzybowski, J.M.V.: Landslide susceptiblity mapping with r.landslide: a free open-source GIS-integrated tool based on Artificial Neural Networks. Environ. Model. Softw. **123**, 104 (2019). https://doi.org/10.1016/j.envsoft.2019.104565

Chen, W., et al.: Landslide susceptibility modeling based on GIS and novel bagging-based kernel logistic regression. Appl. Sci. **2018**(8), 2540 (2018a). https://doi.org/10.3390/app8122540

Chen, W., Pourghasemi, H.R., Zhao, Z.: A GIS based comparative study of Dempster-Shafer, logistic regression, and artificial neural network models for landslide susceptibility mapping. Geocarto Int. **32**, 367 (2016). https://doi.org/10.1080/10106049.2016.1140824

Chen, W., Panahi, M., Pourghasemi, H.R.: Performance evaluation of GIS-based new ensemble data mining techniques of adaptive neuro-fuzzy inference system (ANFIS) with genetic algorithm (GA), differential evolution (DE), and particle swarm optimization (PSO) for landslide spatial modelling. CATENA **157**, 310–324 (2017). https://doi.org/10.1016/j.catena.2017.05.034

Chen, W., et al.: GIS-based landslide susceptibility evaluation using a novel hybrid integration approach of bivariate statistical based random forest method. Catena **164**(2018), 135–149 (2018b). https://doi.org/10.1016/j.catena.2018.01.012

Ebrahimi, E., Solaimani, K., Pourghasemi, H.R.: Application of Geospatial information system and frequency ratio model in landslide sensitivity zoning, Siah Bisheh, Mazandarn, Iran. In: International Conference on Sustainable Development, Strategies and Challenges with a Focus on Agriculture, Natural Resources, Environment and Tourism, Tabriz, Iran, Februrary 2015. (The language of the article is Persian)

Feizizadeh, B., Blaschke, T.: GIS-multicriteria decision analysis for landslide susceptibility mapping: comparing three methods for the Urmia lake basin, Iran. Nat. Hazards **65**(2013), 2105–2128 (2012). https://doi.org/10.1007/s11069-012-0463-3

Feizizadeh, B., Shadman Roodposhti, M., Jankowski, P., Blaschke, T.: A GIS-based extended fuzzy multi-criteria evaluation for landslide susceptibility mapping. Comput. Geosci. **73**, 208 (2014). https://doi.org/10.1016/j.cageo.2014.08.001

Hong, H., Pradhan, B., Xu, C., Tien Bui, D.: Spatial prediction of landslide hazard at the Yihuang area (China) using two-class kernel logistic regression, alternating decision tree and support vector machines. CATENA **133**(2015), 266–281 (2015). https://doi.org/10.1016/j.catena.2015.05.019

Hong, H., Ilia, I., Tsangaratos, P., Chen, W., Xu, C.: A hybrid fuzzy weight of evidence method in landslide susceptibility analysis on the Wuyuan area, China. Geomorphology **290**, 1–16 (2017). https://doi.org/10.1016/j.geomorph.2017.04.002

Jaafari, A., Najafi, A., Pourghasemi, H.R., Rezaeian, J., Sattarian, A.: GIS-based frequency ratio and index of entropy models for landslide susceptibility assessment in the Caspian forest, northern Iran. Int. J. Environ. Sci. Technol. **11**(4), 909–926 (2013). https://doi.org/10.1007/s13762-013-0464-0

Khan, H., Shafique, M., Khan, M.A., Bacha, M.A., Shah, S.U., Calligaris, C.: Landslide susceptibility assessment using frequency ratio, a case study of northern Pakistan. Egypt. J. Remote Sens. Space Sci. **22**(2019), 11–24 (2018). https://doi.org/10.1016/j.ejrs.2018.03.004

Kavzoglu, T., Kutlug Sahin, E., Colkesen, I.: Selecting optimal conditioning factors in shallow translational landslide susceptibility mapping using genetic algorithm. Eng. Geol. **192**(2015), 101–112 (2015). https://doi.org/10.1016/j.enggeo.2015.04.004

Nandi, A., Shakoor, A.: A GIS-based landslide susceptibility evaluation using bivariate and multivariate statistical analyses. Eng. Geol. **110**(2009), 11–20 (2009). https://doi.org/10.1016/j.enggeo.2009.10.001

Ozdemir, A., Altural, T.: A comparative study of frequency ratio, weights of evidence and logistic regression methods for landslide susceptibility mapping: Sultan Mountains, SW Turkey. J. Asian Earth Sci. **64**(2013), 180–197 (2012). https://doi.org/10.1016/j.jseaes.2012.12.014

Rana, P.N., Champati Ray, P.K., Bisht, P., Bagri, D.S., Wasson, R.J., Sundriyal, Y.: Identification of landslide-prone zones in the geomorphically and climatically sensitive Mandakini valley, (central Himalaya), for disaster governance using the Weights of Evidence method. Geomorphology **284**, 41 (2016). https://doi.org/10.1016/j.geomorph.2016.11.008

Peng, L., Niu, R., Huang, B., Wu, X., Zhao, Y., Ye, R.: Landslide susceptibility mapping based on rough set theory and support vector machines: a case of the Three Gorges area, China. Geomorphology **204**(2014), 287–301 (2013). https://doi.org/10.1016/j.geomorph.2013.08.013

Pourghasemi, H.R., Pradhan, B., Gokceoglu, C., Deylami Moezzi, K.: Landslide susceptibility mapping using a spatial multi criteria evaluation model at Haraz Watershed, Iran. Terrigenous Mass Movements **2012**, 23–49 (2012). https://doi.org/10.1007/978-3-642-25495-6_2

Pourghasemi, H.R.: Landslide sensitivity modeling using a hybrid frequency-entropy algorithm, East of Mazandaran province, Iran. In: 11th National Conference on Watershed Management Science and Engineering of Iran, April 2016. (The language of the article is Persian)

Shahabi, H., Hashim, M.: Landslide susceptibility mapping using GIS-based statistical models and Remote sensing data in tropical environment. Sci. Rep. **5**(2015), 9899 (2015). https://doi.org/10.1038/srep09899

Saha, A., Saha, S.: Comparing the efficiency of weight of evidence, support vector machine and their ensemble approaches in landslide susceptibility modelling: a study on Kurseong region of Darjeeling Himalaya, India. Remote Sens. Appl. Soc. Environ. **19**, 100323 (2020). https://doi.org/10.1016/j.rsase.2020.100323

Saeidian, B., Mesgari, M.S., Ghodousi, M.: Evaluation and comparison of genetic algorithm and bees algorithm for location–allocation of earthquake relief centers. Int. J. Disaster Risk Reduction **15**(2016), 94–107 (2016). https://doi.org/10.1016/j.ijdrr.2016.01.002

Singh, S.K., Yadav, S.P.: Intuitionistic fuzzy multi-objective linear programming problem with various membership functions. Ann. Oper. Res. **269**(1–2), 693–707 (2017). https://doi.org/10.1007/s10479-017-2551-y

Pham, B.T., Tien Bui, D., Pourghasemi, H.R., Indra, P., Dholakia, M.B.: Landslide susceptibility assessment in the Uttarakhand area (India) using GIS: a comparison study of prediction capability of naïve bayes, multilayer perceptron neural networks, and functional trees methods. Theoret. Appl. Climatol. **128**(1–2), 255–273 (2015). https://doi.org/10.1007/s00704-015-1702-9

Thai Pham, B., Tien Bui, D., Indra, P., Dholakia, B.: Hybrid integration of Multilayer Perceptron Neural Networks and machine learning ensembles for landslide susceptibility assessment at Himalayan area (India) using GIS. CATENA **149**(2017), 52–63 (2017). https://doi.org/10.1016/j.catena.2016.09.007

Vakhshoori, V., Zare, M.: Landslide susceptibility mapping by comparing weight of evidence, fuzzy logic, and frequency ratio methods. Geomatics. Nat. Hazards Risk **7**(5), 1731–1752 (2016). https://doi.org/10.1080/19475705.2016.1144655

Yalcin, A., Reis, S., Aydinoglu, A.C., Yomralioglu, T.: A GIS-based comparative study of frequency ratio, analytical hierarchy process, bivariate statistics and logistics regression methods for landslide susceptibility mapping in Trabzon, NE Turkey. CATENA **85**(2011), 274–287 (2011). https://doi.org/10.1016/j.catena.2011.01.014

Zadeh, L.A.: Fuzzy sets. Inf. Control **8**(1965), 338–353 (1965). https://doi.org/10.1016/S0019-9958(65)90241-X

Acceleration of Socio-economic Growth of Rural Parts - Nidhal, Khatav A Case Study

Umesh L. Deshpande[1], Shivraj Karape[2], and Anand B. Tapase[3](✉)

[1] Department of Applied Mechanics, Government College of Engineering, Karad, Maharashtra, India
[2] Construction Management, Department of Civil Engineering, Government College of Engineering, Karad, Maharashtra, India
[3] Rayat Shikshan Sanstha's, Karmaveer Bhaurao Patil College of Engineering, Satara, India

Abstract. With the aim of sustainable development of rural parts, the challenges need to be identified so that appropriate solution can be evolved, in the present study Nidhal village from Khatav Taluka is considered as a case study. To identify the problems associated with the basic needs of the rural areas, a preliminary socio-economic technical survey was conducted. Based on the procured data it was observed that the village has problems like solid and sewage waste disposal, water scarcity, poor road network, etc. A development plan based on the identified problems for the village is prepared to evolve at an appropriate solution with the help of available local resources and appropriate technology. The work is decentralized based on the priorities, the design of biodegradable and non-biodegradable waste management system in the form of a detailed project report on solid waste disposal is prepared. For the solution of the solid waste disposal problem, the study and analysis of different methods were carried out. An analytical study for the integrated solid waste management of Nidhal is completed with cost and time analysis. The windrow composting, vermicomposting and landfill methods were selected for Nidhal village. The design, drawings, estimation for all the three methods was prepared. For the solution to excessive charges of drinking water pumping system, solar water pumping system has been designed and installed with the help of MEDA.

Keywords: Sustainable rural development · Wastewater · Sewage water

1 Introduction

India in the twenty-first century is an interesting mosaic of new and old challenges and opportunities. Huge development disconnects such as inequity in health, education, incomes, and assets, opportunities for growth, and access to resources. 70% of the population in India lives in rural areas, engaged in an agrarian economy with agriculture and allied sectors employing 51% of the total workforce but accounting for only 17% of the gross domestic product (GDP). Without rural development, India cannot optimally realize its growth potential and claim its place in the world. To develop the country, we have to start from the villages. So, it is necessary to find out solutions to developmental

A. Tapase et al. (Eds.): GeoChina 2021, SUCI, pp. 89–108, 2021.
https://doi.org/10.1007/978-3-030-79644-0_8

issues in the social field by using research and skilled manpower from these institutes and application of appropriate technology. Unnat Bharat Abhiyan (UBA) is aiming at the participation of higher educational institutions to work with the people of rural India in identifying development challenges and evolving appropriate solutions for accelerating sustainable growth. It is necessary to analyze various social and developmental issues in the state through research and to find out perfect solutions to them. Accordingly, to enhance the research among various Government agencies and educational institutes on serious social and economic problems, it is necessary to boost the participation of students from engineering and other vocational courses in various project-wise schemes.

Socio-economic survey tools are designed to collect information as a means of improving understanding of local resource management systems, resource use, and the relative importance of resources for households and villages. Surveys also provide information on interaction with the government decision-making systems and community perceptions of trends and priority issues. Knowledge about community-based institutions, which is also obtained, and their roles in the sustainable use and conservation of natural resources, helps to facilitate or reinforce a consensus on land tenure and rights for the region, now and in the future. A socio-economic survey is regarded as one of the most important sources of statistical data on household expenditure and income as well as other data on the status of housing, individual and household characteristics, and living conditions.

2 Socio-economic Survey

2.1 Socio-economic Survey of Nidhal Village

A Socio-Economic survey is carried from Nidhal village. The data related to the solid waste management system, sewage treatment system, water supply systems, sources of drinking water, availability of toilets is done by collecting data per the questioner prepared.

2.2 Interpretation of Socio-economic Survey

It is learned from the survey that the waste collection is carried with tractor-trolley daily from the doorsteps of every house which is disposed into the pits as the waste collection treatment is not available as mentioned in Table 1. Sewage water is carried through over and underground gutters into the streams but no treatment plant is available Table 2. Water supply is through Inam, Bormal and old wells and the drinking water is available through tap waters, wells and bore wells Table 3 and Table 4.

From the information given in Table 5 and statistical representation in Fig. 1, it is observed that 98% of the total households of the village have their toilet in their house and 2% of villagers use public toilets. Almost every household has its own toilets.

Table 1. Solid waste management system

Waste collection	Doorstep
Collection schedule	Daily
The vehicle used for the collection	Tractor-trolley
Disposal of waste	Into the Pit
Treatment of collected waste	Not available

Table 2. Sewage water treatment system

Type of Gutter		Existing condition
Kaccha		Average
Pakka	Overground	Good
Pakka	Underground	Good
Treatment Plant		No
Disposal		Directly into Streams

Table 3. Water supply system

Name	Doorstep
Inam Well	TCL powder
Bormal Well	TCL powder
Old Well	TCL powder

Table 4. Source of drinking water

Source of drinking water	Households	Percentage
Tap water	520	72
Well/bore well	200	28
Total	**720**	**100**

Table 5. Availability of Toilet

Availability of Toilet	Households	Percentage
Households with Toilet	703	98
Household without Toilet	17	2
Total	**720**	**100**

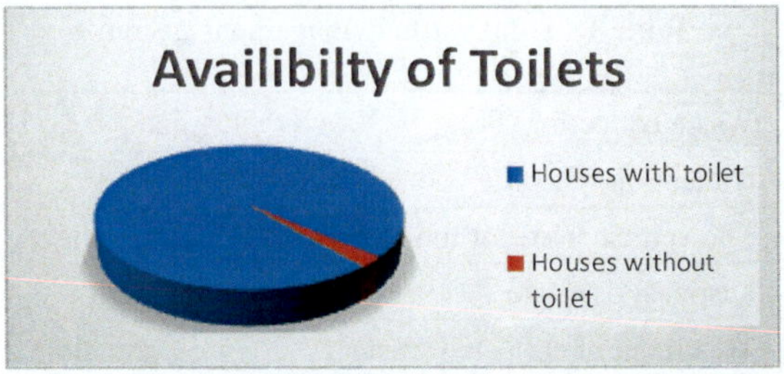

Fig. 1. Statistical representation of availability of toilets

Table 6. Renewable and Non-Renewable energy

Source	Households	Percentage
Renewable Energy	489	42
Non-Renewable Energy	686	58
Total	**1175**	**100**

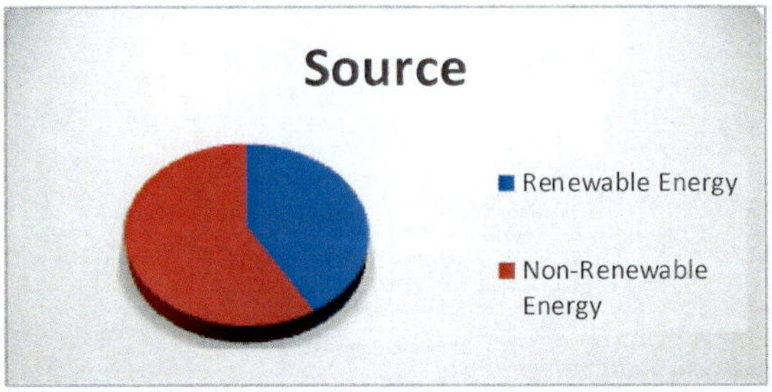

Fig. 2. Sources of energy

From the information shown in Table 6 and from Fig. 2, it is observed that 58% of the total households of the village use renewable energy for cooking and water heating and about 42% of households use non-renewable energy. Due to very less use of renewable energy, the consumption of non-renewable energy is high.

Table 7. Migration status of the village

Migration Status	Population	Percentage
Temporary migrated	467	11
Permanent migrated	96	2
Temporary Reverse Migrated	10	0.2

(*continued*)

Table 7. (*continued*)

Migration Status	Population	Percentage
Permanent Reverse Migrated	31	0.8
Native	3578	86
Total	**4182**	**100**

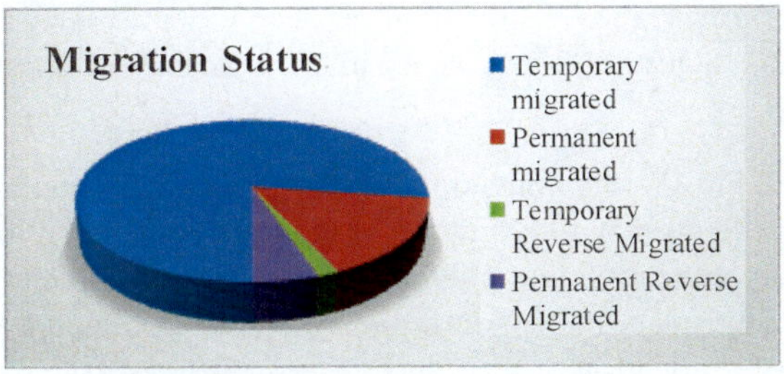

Fig. 3. Migration Status of the total population of the village

From the information shown in Table 7 and Fig. 3, it is observed that 13% of the total population of the village is migrated to different cities in search of a job and also for higher education. From the analysis, it is observed that 1% of the total population of the village is reversed migrated (Table 8).

Table 8. Literacy details of villagers

Literacy	Population	Percentage
Illiterate between age 0–5 year	298	7
Illiterate (remaining)	359	9
Primary school	595	14
Secondary school	1611	38
Higher secondary school	761	18
Graduate	482	12
Postgraduate	76	2

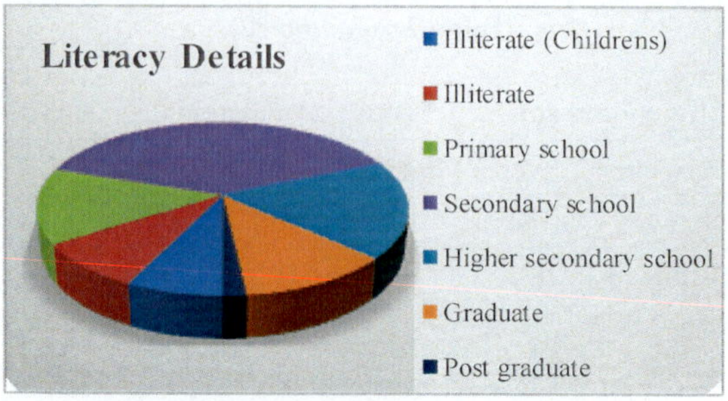

Fig. 4. Statistical representation of literacy details of villagers

From Fig. 8 and Fig. 4, it is noticed that 16% of the total population of the village is illiterate. Those who are not able to write or read are considered illiterate. 38% have completed secondary education and only about 14% have completed their graduation. The percentage of Secondary education in the village is more and only about 2% have completed their post-graduation.

Table 9. Agricultural details

Landholding	Area (Ha)	Percentage
Uncultivable area	100.15	21
Total cultivable area	388.01	79
Total Area	**488.16**	**100**

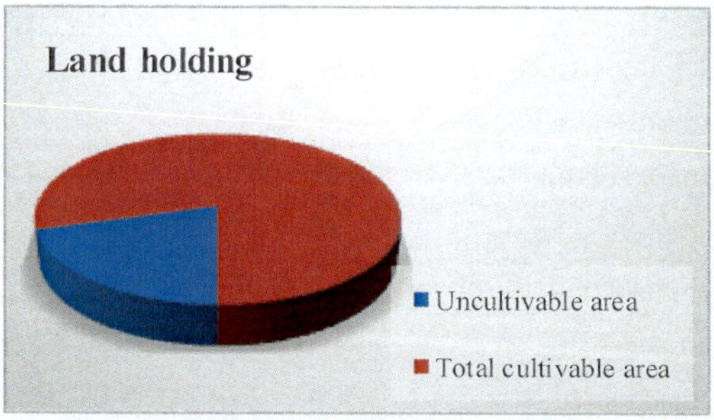

Fig. 5. Statistical representation of landholding

From Table 9 and Fig. 5, it is observed that 79% of the total landholding area of Nidhal village is a cultivable area and 21% of the land is the uncultivable area. This cultivable land is the only source of earning for 16% of the total population of Nidhal village.

2.3 Overall Summary of Socio-economic Survey

a) There is a lack of treatment for solid waste and sewage waste. Sewage waste is directly disposed into streams; also, solid waste is dumped on open ground without any prior treatment.
b) Water is supplied to villagers by using TCL for treatment through the pipeline network. Every day for 70 min at 6 am.
c) Water is supplied through a pipeline network to 72% of households. 28% of households rely on their wells.
d) Roads from the highway to village and village internal roads are of bitumen roads. *Panand* roads of 10 km are in water-bound macadam type, which is about 37.78% of total roads.
e) The water supply system of the village works on MSEB electricity. The electricity cost is about 3.5 Lakhs per annum.
f) All government buildings work on the electricity received from MSEB.
g) There are only 19% of houses built in RCC, remaining all are load-bearing.
h) 98% of houses having their own toilets.
i) 27% population directly depends on agriculture.
j) Only 56% of the total agricultural land is irrigated and the rest 44% of land un-irrigated. 87% of the land is irrigated by using the flooding method.
k) The maximum population falls in the 21–40 age group, which is about 37%.
l) The sex ratio of Nidhal Village is 970 women per 1000 men.
m) 15% of the village population migrated from the village for education and job purpose. Forward migration is more than reverse migration.
n) 84% of the population is literate and 16% is illiterate. From illiterate there are 7% are of below 6th age.
o) Of literate people, 16% of literate can operate a computer.
p) Primary and secondary school facilities are considered modern in the village.
q) 63 houses in the village don't have a gas connection, which is about 9% of the total household.
r) There is disconformity observed between annual income and expenditure of families.

3 Various Methods for Design of the Solid Waste Plant with Financial Analysis and Work Cycle Management

3.1 At Source Reduction and Reuse

The most preferred option for waste management in the ISWM hierarchy is to prevent the generation of waste at various stages including in the design, production, packaging, use, and reuse of products. Waste prevention helps to reduce handling, treatment, and disposal costs and various environmental impacts such as leachate, air emissions, and generation of greenhouse gases (GHG). Minimization of waste generation at the source and reuse of products are the most preferred waste prevention strategies.

3.2 Waste Recycling

Solid Waste Management (SWM) Rules, 2016 defines recycling as "the process of transforming segregated solid waste into a new product or a raw material for producing new products." Further, it also states that "arrangement shall be made to provide segregated recyclable material to the recycling industry through waste pickers or any other agency engaged or authorized by the local body for the purpose"[(10)].

According to the ISWM hierarchy, recycling is a preferred waste management strategy after source reduction and reuse. Recycling systems should be adopted before planning for any waste processing or treatment facilities. It is necessary to recycle waste to recover material resources through segregation, collection, and re-processing to create new products. In the waste management hierarchy, composting is considered an organic material recovery process and is often considered at the same hierarchical level as inorganic waste recycling.

3.3 Composting

Composting is an organic matter that has been decomposed and recycled as a fertilizer and soil amendment and the respective process of production of compost is known as composting. Composed is a key ingredient in organic farming.

Compost is rich in nutrients. It is used in gardens, landscaping, horticulture, and agriculture. The compost itself is beneficial for the land in many ways, including as a soil conditioner, fertilizer, addition of vital humus or humus acids, and natural pesticides for the soil. In an ecosystem, compost is useful for erosion control, land and stream reclamation, wetland construction, and as a landfill cover. Organic ingredients intended for composting can alternatively be used to generate biogas through anaerobic digestion.

3.4 Technological Option for Composting of Organic Waste

1. NADEP Method
2. Bangalore method
3. Indore method
4. Chinese rural composting.
5. Windrow composting
6. In-vessel composting
7. Vermi composting.

3.5 Wastes to Energy

Where material recovery from waste is not possible, energy recovery from waste through the production of heat, electricity, or fuel is preferred. Bio-methanation, waste incineration, production of refuse-derived fuel (RDF), co-processing of combustible non-biodegradable dry fraction from MSW in cement kilns, and pyrolysis or gasification are some waste-to-energy technologies.

3.6 Waste Disposal

Residual inert wastes at the end of the hierarchy are to be disposed of in sanitary lined landfills, which are constructed by stipulations prescribed in SWM Rules, 2016. All over the world, landfills that integrate the capture and use of methane are preferred over landfills which do not capture the landfill gas. As per the hierarchy, the least preferred option is the disposal of waste in open dumpsites. However, Indian laws and rules do not permit disposal of organic matter into sanitary landfills and mandate that only inert rejects (residual waste) from the processing facilities, inert street sweepings, etc. can be landfilled.

4 Solid Waste Management System for Nidhal Village

4.1 General Design Considerations

4.1.1 Population Forecasting

For forecasting of population growth rate: 7.98% per decade is taken. According to that population of Nidhal village will be 4528 in 2041. The number of Households in 2001 was 702 and in 2011 was 791 (Table 10).

Table 10. Population forecasting

Year	Persons	Year	Persons
2001	3330	2031	4193
2011	3596	2035	4393
2021	3883	2041	4528
2025	4040		

4.1.2 Generation of Waste

a. As per CPCB 2014–15 report average generation is 0.11 kg/capita/day.
b. As per the NEERI 1996 report, it is 0.21 kg/capita/day for having a population between 0.1 million to 0.2 million.
c. As per the "Manual on Solid Waste Management" prepared by Central Public Health & Environmental Engineering Organization (CPHEEO), Ministry of Urban Development, Govt. of India, it is 0.2–0.6 kg/capita/day.
d. Waste generation for Nidhal village is considered to be 0.25 kg/capita/day.

4.1.3 Total Generation of SW in Village

- Consider the 2025s population which is $4040 \approx 4050$ persons.
- Generation – $4050 \times 0.25 = 1012.5$ kg/day.
- The total generation will be 1012.5 kg/day.

4.1.4 Physical Composition of Solid Waste

As per "Improving Solid Waste Management in India", Zhu et al. 2008 (Table 11).

Table 11. Physical composition

Composition	1996	2005	Composition	1996	2005
Biodegradables	42.21	47.43	Glass	0.60	1.01
Paper	3.63	8.13	Rags	–	4.49
Plastics & Rubber	0.60	9.22	Other	–	4.016
Metal	0.49	0.5	Inert	45.13	25.16

Above given Composition is for urban areas. We are considering 65% of biodegradable waste for rural areas. Total Generation of Biodegradable waste =1012.5 × 0.60 = 607.5 kg/day ≈ 610 kg/day.

4.2 Detailed Project Report on SWM for Nidhal Village

See Table 12.

Table 12. Details of SW of Nidhal village

Name of the Grampanchayat (GP)	Nidhal	Number of wards	4
Block	Khatav	Total generation of waste	1012.5 kg/day
District	Satara	Total generation of wet waste	607.5 kg/day
Population	3596	Total generation of dry waste	405 kg/day
Households	791	Total generation of recyclable waste	151.87 kg/day

4.2.1 Present Status of Waste Management

For a collection of waste from the village, a very simple but effective method is used. Every day early in the morning, the sanitary workers of Gram Panchayat administration are collecting the waste from each house, i.e. they are using the door-to-door collection system. For collection, Gram Panchayat has a tractor and trolley. The collected waste is not of the segregated one. After collection, they went to a dumping pit having size 7 m × 7 m × 2.5 m. All collected waste is dumped into this pit without any prior treatment on it. After some specific period, the dumped waste is burnt (Fig. 6).

Fig. 6. Solid waste spread on dumping yard

4.2.2 Proposal for Waste Management

4.2.2.1 Community Preparation

The GP residents hold the key to success in solid waste management. The human propensity to respond to a call for any change generally does not receive the same level of cooperation and support from all corners. For taking support from residents it is necessary to generate awareness about health and sanitation. For awareness following points will help.

a. Waste bins distribution with handbills
b. Student Orientation
c. Cultural evening
d. Interpersonal communication
e. SMS alerts
f. Educative information
g. Rangoli Competition
h. Prizes and gifts.

4.2.2.2 Planning of Waste Management

The following diagram shows the waste management process (Fig. 7).

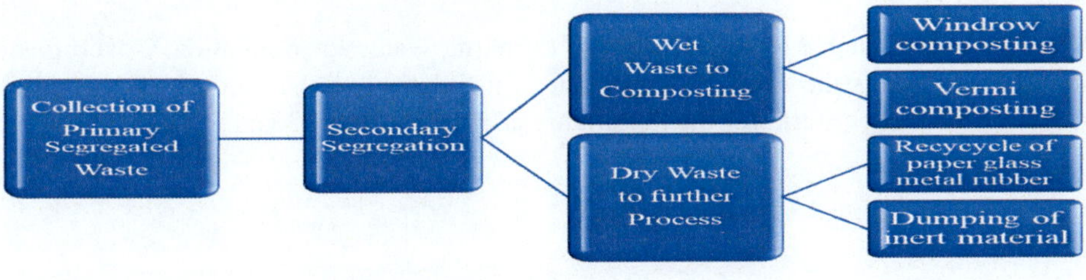

Fig. 7. Flow-chart SWM plan for Nidhal

4.3 Collection of Primary Segregated Waste

Segregation plays an important role in the waste treatment system. It is necessary to segregate the waste at the source of generation. For segregation by stakeholders, the awareness program will help. Also, the village body or Gram Panchayat should distribute 3 buckets to each household. These 3 buckets having the color green for wet waste, blue for dry waste, and red for hazardous waste.

4.4 Secondary Segregation

After the transport of waste to the treatment plant, waste is again segregated by workers. This is especially to ensure that wet waste that will go for vermin or windrow composting does not have anything harmful/mix up of plastic, etc. The recyclable wastes are sort as metal, plastic, rubber, and paper. It is mandatory to achieve 95% to 100% segregation. This function should carry out in 2 h. 4 workers are sufficient for secondary segregation.

4.5 Composting

1. Windrow Composting
2. Vermi Composting

4.6 Some Cares Are Needed While Composting is as Follows

a. As per the schedule given above, it is mandatory to turn the waste periodically.
b. The temperature of the pile should be between 30 °C to 65 °C. If the temperature increases, increase the rate of turning and spread some water on the pile. If temperature decreases, decrease the rate of turning and cover the pile using dry gunny bags.
c. The moisture content should between 40%–60%. If moisture content falls below 40% spread some water on the pile and go above 60% then add some sawdust into a pile and mix it.
d. After the composting period, the compost is feed into a gunny bag and place aside for 1 week. Then send for selling.

4.7 Landfilling

As per guidelines stated by the Ministry of Drinking Water and Sanitation, Government of India, regarding the sanitary landfill, it is mandatory to the landfill of only inert materials. The site criteria for the landfill are as follows (Table 13):

Table 13. Landfill site location criteria

Sr. No.	Landmark	Distance in meter	Sr. No.	Landmark	Distance in meter
1	Lake, pond	200	5	Public park	500
2	River	100	6	Airport	3000 to 20000
3	Highway	500	7	Water supply well	500
4	Habitation	500	8	Groundwater table level	2 m below the base of the landfill

The site should not be in the area of the above-stated criteria. The inert material which is 25% of the total waste should be filled into the landfill. This landfill is designed for 20 years duration. Use excavated material for the cover. The leachate collected into the leachate well should spread on landfill material.

4.8 Land Required for the Composting and Landfill

See Table 14.

Table 14. Area requirement

Sr. No.	Method of composting	Area required (m^2)
1	Windrow Composting	330
2	Vermi Composting	372
3	Landfilling	1935

5 Solar Water Pumping System

5.1 Solar Water Pumping System for Nidhal Village

An estimate of the solar water pumping system was done for the Nidhal Village as mentioned in Table 15.

5.2 Panel Design for Nidhal Solar Water Pumping Systems

There are three open wells in Nidhal village under operation. These three wells are having 2 pumps each of capacity 7 HP and 8 HP. The total electricity bill for water pumping is about 3.5 lakhs per annum. Such extensive expenditure on electricity bills effects on minimizing the overall profit of Gram Panchayat. There is one most suitable alternative for water pumping. A solar water pumping system can be installed in the village. This will reduce the electricity bills as well as such an alternative is eco-friendly. The design of solar panels and stands over which panels are resting is as follows.

Table 15. Estimation for the solar water pumping system

Total project cost	
A. Panel	Rs. 192000
B. Inverter	Rs. 100000
C. Wiring connections	Rs. 60000
D. Stay frame	Rs. 10998 × 4 = Rs. 43992
Sub total	Rs. 395992
Contingency 2%	Rs. 7920
Total	**Rs. 403912**

The solar water pumping system for 8 HP motors requires 20 panels and the total estimated cost is Rs. 4.04 lakhs. This design is carried out for *Inam* well.

5.3 Payback Period Calculations

The life of the plant will be 24 years.
A loan is taken from bank = Rs. 4.04 lakhs.
Interest on loan borrowed = 12% per annum.
Electricity charges = Rs. 1.5 lakhs per annum.
Installment amount = Rs. 1.5 lakh per annum including interest.
The payback period calculations are carried out here.

Table 16. Payback period for solar water pumping system

Year	Amount	Interest Rate (12%)	Remaining Amount
1	Rs. 4,04,000	4,04,000 × 1.12 = Rs. 4,52,480	Rs. 3,02,480
2	Rs. 3,02,480	3,02,480 × 1.12 = Rs. 3,38,778	Rs. 188778
3	Rs. 188778	1,88,778 × 1.12 = Rs. 2,11,430	Rs. 61,431
4	Rs. 61,431	61,431 × 1.12 = Rs. 68,803	Rs. 81,198 (+)

From, Table 16 it shows that the payback period will 3.5 years. Total recovery is of Rs. 4.22 lakhs, against loan of Rs. 4.04 lakhs.

5.4 Electricity Consumption

Capacity of Pump	= 8 HP
Conversion of HP to watt	= 8 × 746 = 5968 W
Daily requirement of electricity	= 5968 × 6 = 35808 watt per day
Yearly requirement of electricity	= 35808 × 365 = 13.07 MW

5.5 Advantage

The solar water pumping system has many advantages that are summarized below:

(i) Low operating cost: No fuel required for the pump like electricity or diesel hence it reduces the operating cost.
(ii) Low maintenance: This technology requires Low-maintenance because of the lack of moving parts
(iii) Harmonious with nature: Another important advantage is that it gives maximum water output when it is most needed i.e. in hot and dry months.
(iv) Flexibility: The panels need not be right beside the well. They can be anywhere up to 20 m away from the well, or anywhere you need the water. These pumps can also be turned on and off as per the requirement, provided the period between two operations is more than 30 s.
(v) The most important advantage is that technology is noiseless.
(vi) It is a non-polluting technology, which means that it does not release greenhouse gases.

5.6 Limitations of PV Pumping System

PV pumping system has its limitations which are as following:

(i) Low yield: Solar pumping is not suitable for very high demand. It gives very low availability of maximum capacity 74.
(ii) Cost: The initial investment cost of this system is high.
(iii) Theft: Theft of solar panels can be a problem in some areas. So, the farmers need to take necessary precautions.
(iv) Area/space: As the efficiency levels are low, the space required is relatively high.
(v) Solar energy is heavily dependent on atmospheric conditions.

6 Results and Discussion About Solid Waste Management System for Nidhal Village

For composting of degradable waste material in Nidhal village, the windrow and the vermicomposting are planned. For comparison time and cost can be taken as parameters. The time and cost required for selected composting methods are different. The waste management system requires some tools and equipment. The cost of these can be considered as an initial investment in tools and equipment. Also, require some recurring investment (operational investment) for day to day works. The comparisons between various components are carried as follows.

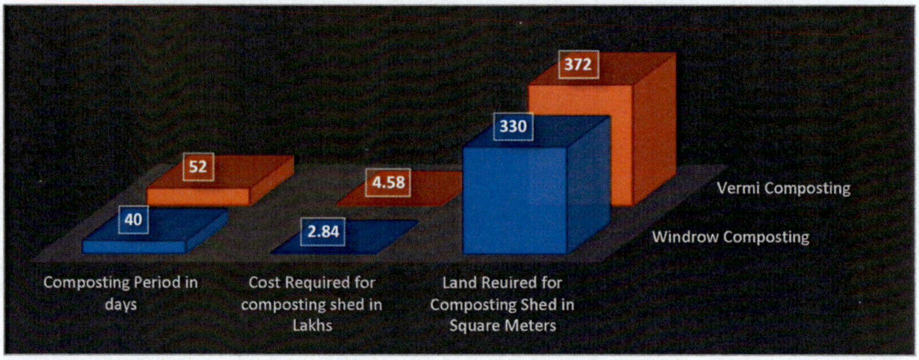

Fig. 8. Comparison of composting methods

6.1 The Time Required for Composting

Figure 8 shows the comparison between the time required for windrow and vermicomposting processes. From Fig. 8 it is observed that the time required for the windrow composting method is 40 days and for vermicomposting is 52 days.

6.2 Cost Required for Composting

Figure 8 shows the cost comparison between the windrow and vermicomposting processes. From Fig. 8 it is seen that the cost required for the windrow composting method is 2.84 lakhs and for vermicomposting is 4.58 lakhs.

6.3 Land Required for Composting Shed

Figure 8 shows the required land comparison of windrow and vermicomposting. From Fig. 8 it is seen that the cost required windrow composting method is 330 m^2 and for vermicomposting is 372 m^2.

After estimation carried out the cost required for the inert material landfill site construction is 17.37 lakhs.

6.4 Solar Water Pumping System for Nidhal Village

The expenditure on the electricity bill is about Rs. 1.5 lakhs per year. Such extensive expenditure on electricity bills effects on minimizing the overall profit of Gram Panchayat. There is one most suitable alternative for water pumping. A solar water pumping system can be installed in the village. This will reduce the electricity bills as well as such an alternative is eco-friendly. For minimizing the cost and give benefits to the village, solar water pumping is designed. The estimated cost for the overall solar water pumping system is Rs. 403912.

7 Conclusions

The Socio-Economical and Technical form was developed by using literature, field experience of previous surveys, and administrative experience. For developing form, help was taken by previous Indo-German socio-economic and watershed management survey form. The time for interaction was chosen with the convenience of the respondents/villagers. Accordingly, the schedule was worked out and the survey was accomplished.

From the survey data, it was observed that any provision to treat solid waste is not available, hence it is directly disposed into the dumping area. Since the solid waste is dumped openly, it is hazardous for villagers. So, the study was carried out to provide an appropriate solution for the treatment and disposal of solid waste.

For solving the solid waste disposal problem, the study was carried out. After a comparative study of various solid waste treatment methods, from the perspective of economy, easy to operate and time required for composting, windrow composting and vermicomposting were found appropriate for Nidhal village. Design and estimation of windrow composting and vermi composting methods for biodegradable material was prepared. As per the design and estimation of both composting methods, it was concluded that the windrow composting method is best suitable for the village under consideration.

Expenses on electricity bills of the water supply system were high. So the study was conducted for finding out an alternative solution with a focus on the non-conventional energy source. The solar water pumping system was found out to be the best alternative for the water supply system. The design of a solar water pumping system was prepared. The proposal for the installation of a solar water pumping system was submitted to a respective Government agency. The MEDAhas accepted the proposal and the tendering process is in progress.

7.1 Based on the Observation of Survey Data Following Conclusions Are Made

1. The collected solid waste is directly dumped to dumping yard without any prior treatment as on date, which may lead to local environmental problems, spreads nuisance and epidemics.
2. The sewage generated is directly disposed to streams without any treatment, this affects the aquatic life.
3. The expenses of electricity required for the drinking water supply system are not affordable.

7.2 Solid Waste Management System

1. The main obstacle in the recycling and disposal problem is the non-segregated waste collection. Waste should be collected after segregation, for better performance of

the designed composting process. So, there should be a separation of waste at the household level as dry and wet waste, afterward the same it collected and transported separately.

2. A combination of vermicomposting with landfill and windrow composting with landfills is the best suitable solution for Nidhal Village. Adopting these methods of treatment, minimize the threat of epidemics and local environmental pollution.
3. The cost required for vermicomposting with landfills is Rs. 2.2 million and for windrow composting with landfills is Rs. 2.02 million. From the above discussion, it is observed that windrow composting with landfills is cheaper than vermicomposting.
4. To complete one cycle of the process for composting, 52 days are required for vermicomposting and 40 days are required for windrow composting. It is observed that 30% more time is required for vermicomposting.
5. Land required for a vermicomposting shed is 372 m^2 and for a windrow composting shed is 330 m^2. It is observed that 13% more land is required for vermicomposting.

7.3 Solar Water Pumping System

1. The overall expenditure on electricity can be minimized, by discontinuing the use of non-renewable energy sources for government building and drinking water pumping systems.
2. The payback period of the solar water pumping system for Nidhal village is 4 years. So after 4th-year expenditure on water pumping system will be negligible. The reduction in expenditure can be utilized for village development.
3. Due to the replacement of energy sources for the drinking water pumping system, a total of 13.07 MW electricity can be saved per year in Nidhal village. This will reduce the stress on electricity demand too.

8 Recommendations

1. Need to establish a wastewater treatment plant in Nidhal village.
2. Necessary to install solar or wind energy systems on all government buildings in the village.
3. For better transport, Panand roads should convert to rigid pavements.
4. Motivate and help villagers to use renewable energy.
5. Necessary to conduct educational workshops for modern agriculture, for better yield in less workforce and expenses.
6. Need to give subsidies for installing drip and sprinkler system for irrigation purpose, helps to minimize loss of water in farms
7. Require to establish agro-based processing industry and giving preference to the villagers in jobs, which will increase jobs and overall economic condition of villagers. Also helps to minimize the rate of migration towards metro cities.
8. Necessary to make awareness about girl childbirth, to raise the sex ratio.
9. The requirement of conducting Praudh Shiksha Abhiyan.
10. Have to provide the gas connection for those who rely on the wood, dung cakes, agro residue, etc.

References

1. Eatwell, J., Milgate, M., Newman, P.N.: Social Economics: The New Palgrave. Macmillan Press Ltd., New York (1989)
2. Bhola, P.: Importance of Socio-Economic Survey. Int. J. Res. Publ. Seminar **5**(2), 23 (2014)
3. "Iraq Household Socio-Economic Survey Manual", Organisation for Statistics and Information Technology, Ministry of Planning, Government of Iraq
4. "Socio-Economic Survey Form", National Sample Survey Office, Government of India, Ministry of Statistics
5. Joshi, R., Ahmed, S.: Status and challenges of municipal solid waste management in India: a review. Cogent Environ. Sci. **2**(1), 1139434 (2016)
6. Soni, A., Patil, D., Argade, K.: Municipal solid waste management. In: Procedia Environmental Sciences, pp. 119–126 (2016)
7. Ahsan, A., Alamgir, M., El-Sergany, M.M., Shams, S., Rowshon, M.K., Nik Daud, N.N.: Assessment of municipal solid waste management system in developing country. Hindawi Publ. Corporation Chin. J. Eng. **12**, 1–11 (2014)
8. Sethi, S., Kothiyal, N.C., Nema, A.K., Kaushik, M.K.: Characterization of municipal solid waste in Jalandhar City, Punjab, India. J. Hazard. Toxic Radioactive Waste © ASCE, pp. 97–106 (2013)
9. Swachh Bharat Mission "Municipal Solid Waste Management Manual", Central Public Health and Environmental Engineering Organization (CPHEEO). www.swachhbharaturban.gov.in
10. Center for Rural Infrastructure and National Institute of Rural Development and Panchayti Raj, "Solid Waste Management in Rural Areas Manual". www.nird.org.in
11. Swachh Bharat Mission (Gramin) "Technological Option for Solid Waste Management in Rural Areas", Ministry of Drinking Water and Sanitation Govt. of India. www.mdws.nic.in
12. Malwana, C., Weerasinghe, T.K., Pilapitiya, S.: Determination of optimal pile dimensions during thermophilic windrow composting of municipal solid waste (MSW) in SriLanka. Int. J. Biosci. Biochem. Bioinf. **3**, 552–556 (2013)
13. Kuhlman, L.R.: Windrow composting of agricultural and municipal waste. In: Resources, Conservation and Recycling, vol. 4, pp. 151–160. Elsevier Science Publishers (1990)
14. Katre, N.H.: Use of vegetable waste through aerobic composting of village Bamhani, District-Gondia. Int. J. Life Sci. Biotechnol. Pharma Res. **1**(4), 134–142 (2012)
15. Eberhard Griineklee, C.: Comparing Open versus In-Vessel composting. In: Asia-North-American, Waste Management Conference, pp. 68–90, December 1998
16. Ambade, B., Sharma, S., Sharma, Y., Sharma, Y.: Assessment of windrow composting plant's performance at Keru, Jodhpur. India. Asian J. Environ. Sci. **6**, 107–111 (2011)
17. Mengistu, T., Gebrekidan, H., Kibret, K., Woldetsadik, K., Shimelis, B., Yadav, H.: Comparative effectiveness of different composting methods on the stabilization, maturation, and sanitization of municipal organic solid wastes and dried fecal sludge mixture. Environ. Syst. Res. **6**(1), 1–16 (2017). https://doi.org/10.1186/s40068-017-0079-4
18. Vinod Kumar, G., Saravana Kumar, V.R., Rama Krishnan, S., Elango, A., Edwin, S.C.: Windrow composting as an option for disposal and utilization of dead birds. Veterinary World **7**, 377–379 (2014)
19. Nagavallemma, K.P., et al.: Vermicomposting: recycling wastes into valuable organic fertilizer. e-J. ICRISAT **2**(1), 2432 (2006)
20. Albasha, M.O., Gupta, P., Ramteke, P.W.: Management of kitchen waste by vermicomposting using earthworm, Eudrilus Eugenia. In: International Conference on Advances in Agricultural, Biological & Environmental Sciences, pp. 81–84, July 2015
21. Pandit, N.P., Ahmad, N., Maheshwari, S.K.: Vermicomposting biotechnology: an eco-loving approach for recycling of solid organic wastes into valuable biofertilizers. J. Biofertil. Biopest. **3**(1), 1–8 (2012)

22. Nagarajappa, D.P., Rashmi, B.N.: Engineering landfill design for municipal solid waste management, Bangalore. Int. J. Eng. Res. Technol. **4**(2), 1117–1120 (2015)
23. Yedla, S.: Modified landfill design for sustainable waste management. Int. J. Glob. Energy Issues **23**(1), 93–95 (2005)
24. Ayub, S., Khan, A.H.: Landfill practice in India: a review. J. Chem. Pharm. Res. **3**(4), 270–279 (2011)
25. Khajuria, A., Yamamoto, Y., Morioka, T.: Estimation of municipal solid waste generation and landfill area in Asian developing countries. J. Environ. Biol. **31**(5), 649–654 (2010)
26. Haruna, R.L., Alaga, T.A., Gajere, E.N., Chioma, U., Amos, S.I.: Landfill site selection for solid waste management in KaruLga, Nasarawa State, Nigeria. Int. J. Trend Res. Dev. **3**(6), 436–467 (2016)
27. "Guidelines for Design and Operation of Municipal Solid Waste Landfills in Tropical Climates" International Solid Waste Association, February 2013
28. Dutta, R.K., Gayathri, V.: Landfill Planning and Design Considerations. Research Gate Publications, January 2012
29. Khare, V., Nema, S., Baredar, P.: Status of solar and wind renewable energy in India. In: Renewable and Sustainable Energy Reviews Published by Elsevier, pp. 1–10 (2013)
30. Gadekar, S.R., Deshpande, P.A., Inamdar, A.R.: Review on solar water pumping system. In: 4th International Conference on a Recent Conference in Science Engineering and Management India International Center, New Delhi (ICRISEM-16), 20th March 2016
31. Campana, P.E., Li, H., Zhang, J., Zhang, R., Liu, J., Yan, J.: Economic optimization of photovoltaic water pumping systems for irrigation. In: Energy Conservation and Management, pp. 32–41 (2015)
32. Korpalea, V.S., Kokatea, D.H., Deshmukha, S.P.: Performance assessment of solar agricultural water pumping system. Int. J. Current Microbiol. Appl. Sci. **7**(5), 133–142 (2018)
33. Maurya, V.N., Ogubazghi, G., Misra, B.P., Maurya, A.K., Arora, D.K.: Scope and review of photovoltaic solar water pumping system as a sustainable solution enhancing water use efficiency in irrigation. Am. J. Biol. Environ. Stat. **1**(1), 1–8 (2015)
34. Eker, B.: Solar-powered water pumping systems. Trakia Univ. Trakia J. Sci. **3**(7), 7–11 (2010)
35. Singh, B., Mishra, A.K.: Utilization of solar energy for driving a water pumping system. Int. Res. J. Eng. Technol. (IRJET) **2**(3), 1284 (2015)
36. Chandel, S.S., Chandel, R.: Review of solar photovoltaic water pumping system technology for irrigation and community drinking water supplies, vol. 49, pp. 1084–1099, September 2015
37. Lu, S., Pei, L., Bai, X.: Study on PV water pumping for different application's. In: 2011 International Conference on Energy, Automation and Signal (ICEAS), 28–30 December 2011
38. Tiwari, A.K., Kalamkar, V.R.: Effects of total head and solar radiation on the performance of solar water pumping system. In: Renewable Energy (2018)
39. Approach and Methodology for Socio-Economic Survey: Baseline and End-line, JRI Research, World Bank Document (2014)
40. "Sindh Union Council and Community Economic Strengthening Support Program", EU-RSPN (2016)

Assessing the Bearing Capacity of Backfills by Stress Wave Velocity and Cone Penetration Resistance

Jiunnren Lai[1(✉)], Ming-Hong Lai[1], and Chiung-Fen Cheng[2]

[1] Department of Construction Engineering, Chaoyang University of Technology, Taichung City, Taiwan
{jrlai,s10911602}@cyut.edu.tw

[2] Department of Multimedia and Game Design, Overseas Chinese University, Taichung City, Taiwan
cfcheng@ocu.edu.tw

Abstract. Sand-cone test is one of the most common tests performed to assess the compaction quality of backfills for earth works. However, performing such test is time consuming and can only obtain the density and water content of backfills. In order to obtain their engineering properties, such as strength or bearing capacity, additional tests such as California Bearing Ratio (CBR) test or plate loading test must be performed. In this study, a miniature dynamic cone penetrometer was developed. The feasibility of assessing the bearing capacity of backfills by stress wave velocity and dynamic cone penetration resistance was investigated. Samples of a silty clay were compacted with various compaction energy and moisture contents. In addition to the density and moisture content, the CBR, P-wave and S-wave velocity, as well as penetration resistance of a mini cone of these samples were measured. Linear relationships between the CBR and S-wave velocity/penetration resistance were found. Therefore, it is feasible to assess the bearing capacity of backfills by these two techniques.

1 Introduction

In engineering practices, the sand-cone test (ASTM 2007) is usually performed to assess the compaction quality of pavement subgrade, pavement bases, pavement sub-bases, embankments, retaining wall backfills, etc. However, the sand cone test is a destructive test and performing this test is labor intensive and time consuming. Alternatively, a neutron probe can be used for rapid determination of moisture content and density of soil and soil-aggregate (ASTM 2010). However, potential health risks to the operator, and the costs of complying with the special requirements for operation and storage of the nuclear gauges, prevent them from been used widely.

Several investigators have been devoted their efforts to develop a more economical and efficient technique to assess the compaction quality of backfills. For examples, Yu and Drnevich (2004) proposed using a time domain reflectometry (TDR) probe to measure the gravitational water content and dry density of compacted soils. Dynamic cone penetrometer (DCP) was adapted for soil compaction determination at shallow

A. Tapase et al. (Eds.): GeoChina 2021, SUCI, pp. 109–116, 2021.
https://doi.org/10.1007/978-3-030-79644-0_9

depths (ASTM 2008). Briaud et al. (2009) developed a new instrument called Briaud Compaction Device (BCD) to measure the soil modulus near the ground surface. Commuri et al. (2009) calibrated an intelligent asphalt compaction analyzer (IACA) based on neural network technology to measure the density of an asphalt pavement continuously during its construction in real time. Lai et al. (2011) investigated the feasibility of evaluating the compaction quality of backfills by stress wave velocities. Kim et al. (2010) used both DCP and Clegg hammer tests for the quality control of roadway compaction. The American Society of Agricultural and Biological Engineers (ASABE 2013) recommended a soil cone penetrometer as a measuring device for characterizing the penetration resistance of soils. The force required to press the circular cone through the soil, expressed in kilopascals, is an index of soil strength called the cone index. Donohue et al. (2013) and Ni et al. (2018) evaluated the compaction quality of backfills by measuring their shear wave velocity using multi-channel analysis of surface wave (MASW) test. Based on the above literature reviews, it seems that both the DCP and the stress wave velocity measurement are all very suitable as alternatives to the sand cone test for the assessment of the compaction quality of backfills.

2 Methodology

The objective of this paper is to compare the advantages and disadvantages between the DCP test and stress wave velocity measurement in assessing the compaction quality of backfills. Because the ASTM DCP is too big to be used for testing samples prepared with a standard compaction mold, a down-sized miniature dynamic cone penetrometer was developed for the use of this study. As shown in Fig. 1, the miniature DCP has a drop weight of 3.92 N (400 g) and a falling distance of 305 mm (12 in.), with a cone tip angle of 30°.

A silty clay of low plasticity was used as testing material for the comparison. The silty clay contains 18% of fine (particle size < 0.075 mm), and has a liquid limit and plasticity index of 24% and 7%, respectively. It is classified as a silty clay with sand (CL-ML) according to the Unified Soil Classification (UCS) system.

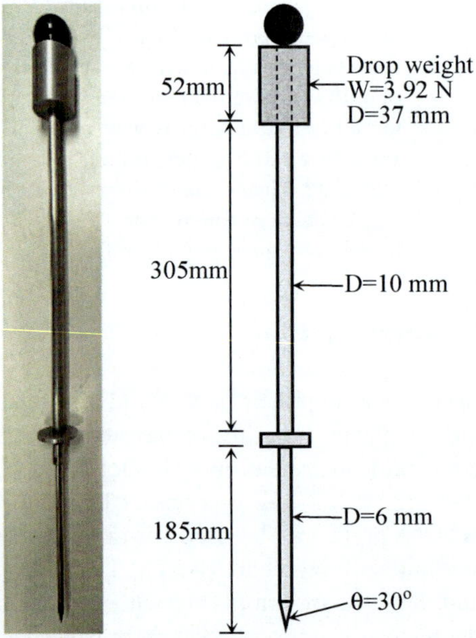

Fig. 1. Schematic drawing of the miniature DCP

A specimen was made by compacting the silty clay in 5 layers with a modified Proctor hammer following the ASTM D1557 Method-C standard procedures. Three different moisture contents (7.5, 10, and 12.5%) and compaction energy (37, 56, and 75 blows per layer) were used in preparing the specimens. Because the stress wave velocity measurement is non-destructive, after a specimen was compacted, the apparatus shown in Fig. 2 (Lai et al. 2011) was used to measure both the P-wave and the S-Wave velocities of the specimen at first. Then, the ASTM D1833 standard procedures were followed to measure the California Bearing Ratio (CBR) on the top surface of the specimen. The miniature DCP was then used to measure the penetration resistance at four different locations on the bottom surface of that specimen. Finally, the soil was extruded out and the weight and moisture content of that specimen were measured.

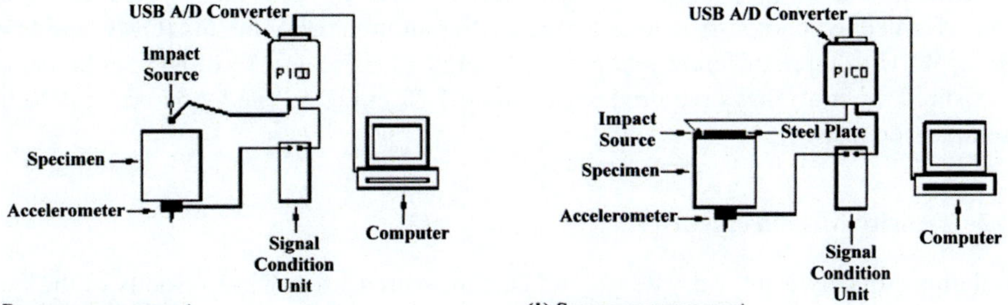

(*a*) P-wave measurement (*b*) S-wave measurement

Fig. 2. Schematic drawing of the test setup for stress wave velocity measurement

3 Results and Discussion

3.1 Penetration Resistance Determination

Two different types of penetration resistance are usually used in a DCP test – the penetration depth after a certain number of blows, or the # of blows required for the cone to penetrate a certain depth. For field measurements, it would be more convenient to use the later. In order to obtain an optimum penetration depth and the effects of testing location, a series of DCP tests were performed on the top and bottom surface of compacted specimens and the penetration depth after 5 and 10blows were recorded. Results of these tests are shown in Fig. 3.

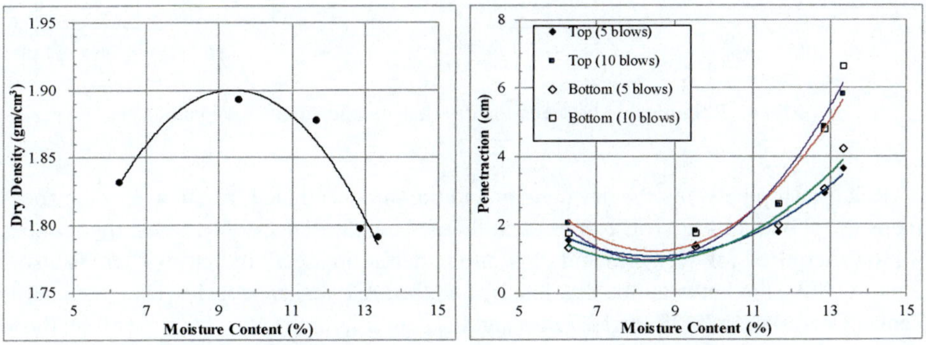

(a) Compaction curve (b) Penetration resistance

Fig. 3. Results of trial penetration resistance tests

It can be seen from Fig. 3a, the silty clay has a maximum dry density of about 1.90 g/cm^3, and an optimum moisture content of about 10%. As shown in Fig. 3b, the penetration depth increases nonlinearly as the moisture content increases for the two different blow counts. Furthermore, for a given moisture content under the same blow count, the penetration depth of the bottom surface is lower that of the top surface and the difference is smaller near the OMC.

Based on these preliminary test results, it can be seen that, except for the very wet specimen, most of the penetration depths fall between 1–3 cm. Therefore, it is decided to performed the comparison tests at three different moisture contents (OMC and OMC ± 2.5%) with three different compaction energy (37, 56, and 75 blows per layer) and measured the # of blows required to penetrate 1.27 cm (0.5″) and 2.54 cm (1″) for the comparison tests.

3.2 Density Measurement

Relationship between the dry density and the measured stress wave velocity of the specimens are shown in Fig. 4. It can be seen that both the P-wave velocity (V_p) and the S-wave velocity (V_s) increase as the dry density increases. There is a linear relationship between the dry density and the stress wave velocity. However, the coefficients of correlation are not very high (0.34 for V_p and 0.35 for V_s). Since the stress wave velocity is a function of density, moisture content, as well as soil type rather than density along, a low coefficient of correlation thus can be expected. This observation is in accord with the previous finding (Lai et al. 2011).

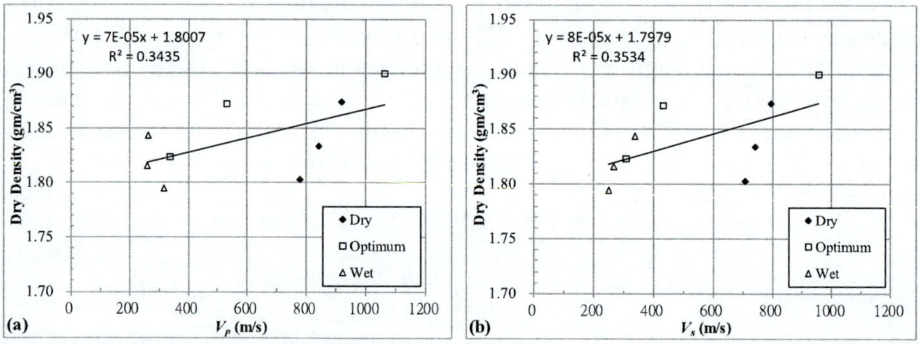

Fig. 4. Relationship between the dry density and stress wave velocity

Relationship between the dry density and the measured DCP resistance of the specimens are shown in Fig. 5. It can be seen that, as the dry density increases, the number of blows required for 12.7 mm and 25.4 mm penetration are all increase. There is also a linear relationship between the dry density and the DCP resistance. However, the coefficients of correlation (0.23 for 12.7 mm and 0.21 for 25.4 mm) are even lower than those of wave velocity measurement.

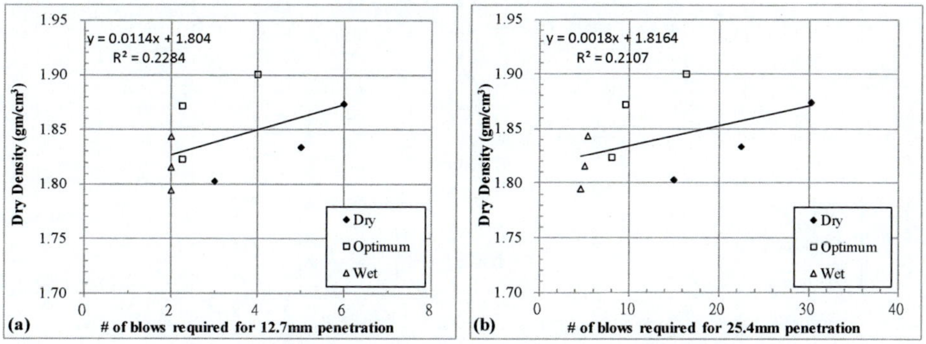

Fig. 5. Relationship between the dry density and DCP resistance

3.3 Moisture Content Measurement

Relationship between the moisture content and the measured stress wave velocity of the specimens are shown in Fig. 6. It can be seen that both the V_p and the V_s increase as the moisture decreases. There is also a linear relationship between the moisture content and the stress wave velocity. The coefficients of correlation are 0.61 for the V_p and 0.58 for the V_s, respectively. These coefficients of correlation are a little better than those of the dry density and stress wave velocity correlation. It is thus concluded that the moisture content plays a more important role in wave velocity than does the dry density.

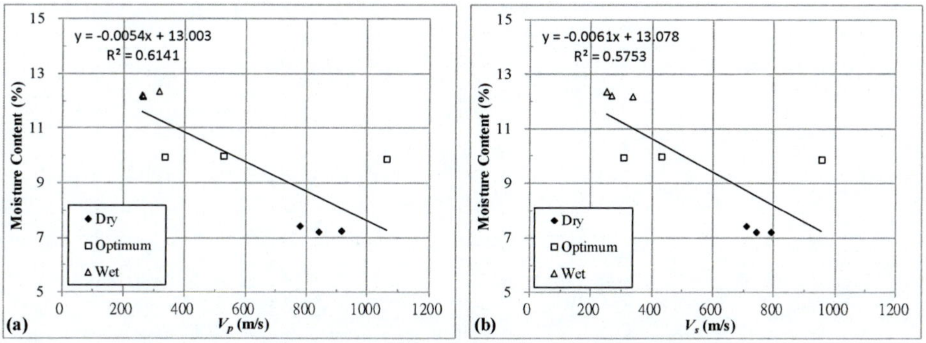

Fig. 6. Relationship between the moisture content and stress wave velocity

Relationship between the moisture content and the measured DCP resistance of the specimens are shown in Fig. 7. It can be seen that, the # of blows required for 12.7 mm and 25.4 mm penetration all increase as the moisture content decreases. There is also a linear relationship between the moisture content and the DCP resistance. In contrast, the coefficients of correlation (0.63 for 12.7 mm and 0.77 for 25.4 mm) are higher than those of wave velocity measurement. Therefore, the DCP resistant performs better in moisture content determination than does the stress wave velocity. Furthermore, measuring the # of blows required for 25.4 mm penetration is better than for 12.7 mm penetration.

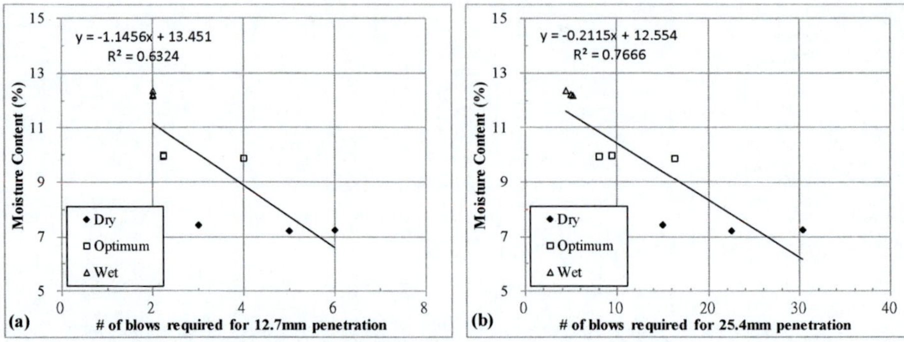

Fig. 7. Relationship between the moisture content and DCP resistance

3.4 California Bearing Ratio Measurement

Relationship between the CBR and the measured stress wave velocity of the specimens are shown in Fig. 8. It can be seen that both the V_p and the V_s increase as the CBR increases. A very good linear relationship between the CBR and the stress wave velocity is obtained as indicated by the high coefficients of correlation (about 0.97 for both the V_p and the V_s). Since the CBR and the stress wave velocity are all overall result of the summation effects of the type, density, moisture content, … of soils. A good linear relationship between the CBR and the stress wave velocity thus can be expected. This observation is also in good agreement with the previous study (Lai et al. 2011).

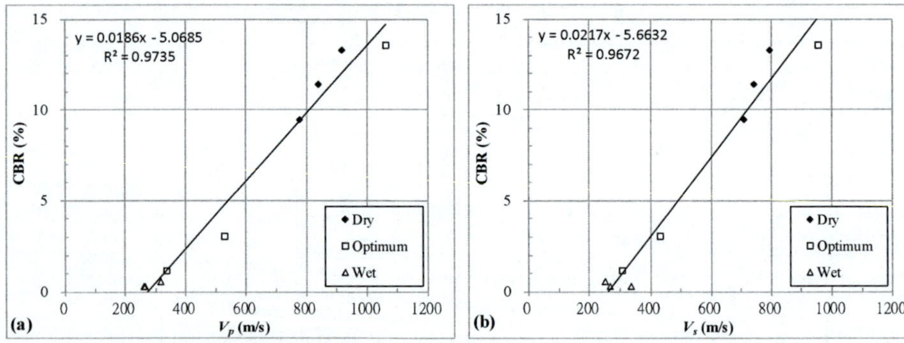

Fig. 8. Relationship between the CBR and stress wave velocity

Relationship between the CBR and the measured DCP resistance of the specimens are shown in Fig. 9. It can be seen that the DCP resistance increases as the CBR increases. There is a relatively good linear relationship between the moisture content and the DCP resistance with coefficients of correlation around 0.8. The coefficient of correlation for 25.4 mm (0.81) penetration is again slightly higher than that of 12.7 mm penetration (0.80), which means a penetration depth of 25.4 mm is more suitable for compaction quality control application.

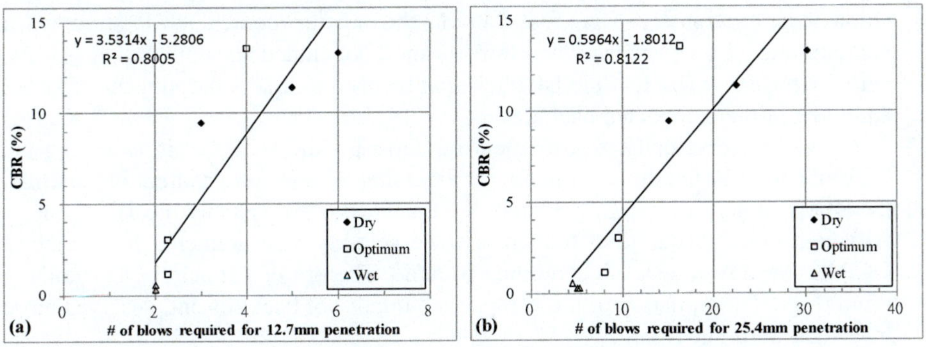

Fig. 9. Relationship between the CBR and DCP resistance

3.5 Correlation between Stress Wave Velocity and DCP Resistance

Relationship between the measured stress wave velocity and the DCP at 25.4mm of the specimens are shown in Fig. 10. It can be seen that both the V_p and the V_s increase linearly as the DCP resistance increases with a fairly good coefficients of correlation around 0.7. Therefore, in addition to the moisture content and CBR, the DCP test can also be used to evaluate the stress wave velocity of compacted soils.

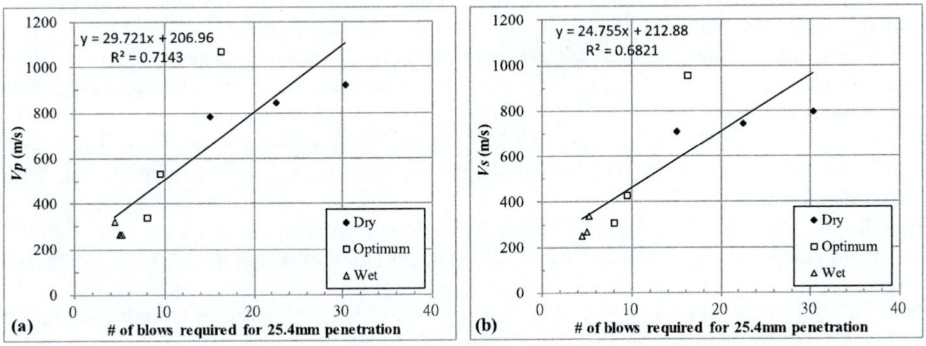

Fig. 10. Relationship between the stress wave velocity and DCP resistance

4 Summary and Conclusions

Both the DCP resistance and the stress wave velocity measurement have long been used for the assessment of the compaction quality of earthworks. In this study, a miniature dynamic cone penetrometer was developed. A series of compaction tests with measurements of CBR, stress wave velocity, as well as DCP resistance were performed. The advantages and disadvantages between stress wave velocity and DCP resistance in assessing the compaction quality of backfills are compared. Based on the results of these tests, the following conclusions can be drawn:

1. Good linear relationships were found among the moisture content, CBR, stress wave velocity, and DCP resistance. Therefore, it can be concluded that both the stress wave velocity measurement and the DCP test can be used to assess the moisture content and CBR of the compacted backfills.
2. The stress wave velocity measurement outperforms the DCP test in assessing the CBR of backfills. On the other hand, for assessing the moisture content of backfills, the DCP test performs slightly better than the stress wave velocity measurement.
3. Although not as accurate as the stress wave velocity measurement, the miniature DCP has the advantages of economic and rapid. Therefore, it should be a promising tool for the determination of the compaction quality of backfills and deserves more investigation in the future.

References

ASAE standard S313.3: Soil Cone Penetrometer. The American Society of Agricultural and Biological Engineers, St. Joseph, MI (2013)

ASTM Standard D1883-99: Standard Test Method for CBR (California Bearing Ratio) of Laboratory-Compacted Soils. ASTM International, West Conshohocken, PA (1999)

ASTM Standard D1557-00: Standard Test Methods for Laboratory Compaction Characteristics of Soil Using Modified Effort (56,000 ft-lbf/ft^3 (2,700 kN-m/m^3)). ASTM International, West Conshohocken, PA (2000)

ASTM Standard D1556-07: Standard Test Method for Density and Unit Weight of Soil in Place by the Sand-Cone Method. ASTM International, West Conshohocken, PA (2007)

ASTM Standard D7380-08: Standard Method for Soil Compaction Determination at Shallow Depths Using 5-lb (2.3 kg) Dynamic Cone Penetrometer. ASTM International, West Conshohocken, PA (2008)

ASTM Standard D6938-10: Standard Test Method for In-Place Density and Water Content of Soil and Soil-Aggregate by Nuclear Methods (Shallow Depth). ASTM International, West Conshohocken, PA (2010)

Briaud, J. L., Rhee, K. Y., and Saez, D.: The BCD: A New Instrument for Compaction Control. Final Report for Highway IDEA Project 118, Transportation Research Board, Washington, DC (2009)

Commuri, S., Mai, A., Zaman, M.: Calibration procedures for the intelligent asphalt compaction analyzer. J. Test. Eval. **37**(5), 454–462 (2009)

Donohue, S., Forristal, D., Donohue, L.A., L.A. : Detection of soil compaction using seismic surface waves. Soil Tillage Res. **128**, 54–60 (2013)

Kim, H., Prezzi, M., Salgado, R.: Use of Dynamic Cone Penetration and Clegg Hammer Tests for Quality Control of Roadway Compaction and Construction. Publication FHWA/IN/JTRP-2010/27. Joint Transportation Research Program, Indiana Department of Transportation and Purdue University, West Lafayette, IN (2010)

Lai, J., Wu, S.M., Chiang, C.H.: Evaluating the compaction quality of backfills by stress wave velocities. J. Test. Eval. **39**(5), 785–791 (2011)

Yu, X., Drnevich, V.P.: Soil water content and dry density by time domain reflectometry. J. Geotech. Geoenv. Eng. **130**, 922–934 (2004)

Ni, S.H., Chang, W.J., Yang, Y.Z., Fan, E.S.: Evaluating Compaction Quality during Earth Dam Construction Using Multi-Channel Analysis of Surface Wave. Geotechnical Earthquake Engineering and Soil Dynamics V: Slope Stability and Landslides, Laboratory Testing, and In Situ Testing, ASCE GSP 293, pp. 432–442 (2018)

Potential of Fired Clay Brick for Use as Short Beams and Columns

Mustapha Mohammed Alhaji(✉), Musa Alhassan, Taiye Waheed Adejumo, Perpetus Chukwuma Ibe, and Mohammed Shehu

Department of Civil Engineering, Federal University of Technology, Minna, Nigeria
a.mustapha@futminna.edu.ng

Abstract. A clay soil, collected from Paggo village, along Minna-Paiko road on Niger state, Nigeria was, molded into clay bricks and fired at temperature of 600 °C. The molds were formed specially with grooves and protrusions of varied depths and thicknesses on both sides. The molds with groves and protrusions were separately used to cast the various bricks. The bricks with protrusions were then fitted into those with grooves using cement slurry of 0.7 water-cement ratio to form short beams. These arrangements categorized the interlocking bricks into three groups (A, B and C), based on depth and thickness of the protrusions and grooves. The results indicated elements formed, interlocking clay bricks of category B, which has higher space between the grooves and protrusions, for cement slurry binder, gave highest compressive strength of 3.7 N/mm^2, which satisfy the strength for load bearing walls according to Nigerian Industrial Standard. Also, elements formed, interlocking clay bricks of category A, which has longer and deeper protrusions and grooves respectively, gave the highest flexural strength of 0.48 N/mm^2, which is far above the near zero flexural strength attributed to masonry clay bricks. This study showed that incorporating relatively longer and deeper protrusions and grooves respectively, as interlocks of fired clay bricks will make withstand reasonable flexural strength that can make the elements serve as short beams.

1 Introduction

The binding properties of cement, which increase in the presence of moisture, makes it took center stage in building and construction industries, since its discovery in 19th century. The setbacks associated with its production, such as emission of carbon dioxide into the atmosphere, high cost of energy, coupled with its relative high cost later became issues of concern, thereby making its use in low cost buildings economically unwise. The cost of cement, which accounts for major cost of development of basic infrastructural, such as roads, water supply, hospitals, schools, houses and ports have tremendously escalated due to hike in cost of energy, required in its production. The carbon dioxide emitted during its production is also hazardous to environment and is therefore needed to be minimized. This has led to various researches on comparative cost analysis of various ways of using cement in building construction industry. Ogunbiyi *et al.* [1] carried out study to compare clay brick and sandcrete blocks as building components.

A. Tapase et al. (Eds.): GeoChina 2021, SUCI, pp. 117–131, 2021.
https://doi.org/10.1007/978-3-030-79644-0_10

The result showed that cost of sandcrete blocks ranges between 120 to 150 naira, while the stabilized clay brick ranges between 45 to 60 Naira. The hike in the price of cement is a major concern to Nigerian government especially to its target of *housing for all by the year 2020* [2].

The developments in construction industry are based on innovative ideas. Addition of other materials to reduce the use of cement is one of the initiatives [3]. A recent research used clay soil as partial replacement for cement as a binder for concrete and mortar production. Bentonite clay was added in order to reduce the percentage of cement in concrete without changing its properties. The mixture was made with different proportions and the highest strength was achieved at 5% addition of Bentonite clay. The use of clay as building material has been considered with upmost importance because of its availability and cost effectiveness [4].

Unfired clay bricks without any form of stabilizer have been used successfully to produce clay bricks for building constructions. Kiptum *et al.* [5] work on effect of depth of clay sample on its suitability for clay brick production. The amount of clay brick lost per annum was also evaluated using peoples' opinion. The author concluded that clay soil samples below the depth of 0.5 m was observed to be suitable for clay brick production and was recommended to avoid loss of brick during processing and handling. Incorporation of natural fibers in production of unfired compressed clay bricks was studied by Njau and Park [6]. Compressive strength of both the fiber reinforced and unreinforced clay bricks were observed to range from 1.74 to 6.85 N/mm^2. Field tests required to evaluate suitability of clay soils for use in production of clay bricks was studied by Suryakanta [7], using smearing test, ball test, wet ball test, dry ball test, sedimentation test, soil shape test and lime tests. The author highlighted the range of parameters that are required to be satisfied by clay before it can be suitably used for clay brick production. Some of these parameters are liquid limit to range from 25 to 38%; plasticity index to range from 7 to 16%; volumetric shrinkage to range from 15 to 25%; composition of sand to range from 20 to 45%; composition of silt to range from 25 to 45% and composition of clay to range from 20 to 35%. According to Murali *et al.* [8], it is essential to identify the characteristics of a clay soil before any construction activities are carryout with it.

Clay materials are sometimes modified or stabilized to manufacture clay bricks so as to enhance their strength and durability. Alumina filler waste and coal ash has been used along with some proportion of clay to produce unfired clay bricks [9]. Unconfined Compressive Strength (UCS) was observed to increase from minimum of about 5 N/mm^2 to over 28 N/mm^2, depending on composition of the admixtures. However, the strength was found to decrease with increase in clay content. Venkatesan [10] studied bricks made of clay, mixed with wood ash and charcoal and observed that at 15% charcoal ash, the bricks have compressive strength of 11.07 N/mm^2, while the maximum strength for those with 15% wood ash was 7.9 N/mm^2. Comparative study on the performance of normal clay bricks and those made using fly ash stabilized clay was carried out by Kumar and Hooda [11]. It was observed that the weight of fly ash clay stabilized bricks reduced by 29% compared to the normal clay bricks and the strength is higher than the best carrying capacity of normal clays by 25%. Chokshi *et al.* [12] carry out a similar study to compare normal clay bricks and fly ash stabilized clay bricks in Gujarat region of India.

The author also reported that fly ash bricks are far lighter than the normal clay bricks and possesses higher strength. Bricks made of clay, replaced with varied composition of materials ranging from cement, sawdust, lime and sand was considered by Smeu *et al.* [13]. The compressive strength was observed to be higher for mixtures containing lime compared to the mixtures containing cement alone. The mixtures with lower water value of 15% gave higher strength compared to those with 25% water content. The flexural strength ranges from 0.4 to 2.6 N/mm^2, while the compressive strength ranges from 2.75 to 6.2 N/mm^2.

Treatment of soil with cement reduces the volume change, but this type of treatment becomes unsuitable for soils with high plasticity index [14]. Soil can be stabilized by addition of cement or lime. Such stabilization processes improve various engineering properties of the soil resulting to improved construction materials. Increase in soil strength, durability, stiffness and reduction in plasticity and swelling/shrinkage potential are some of the benefits of soil stabilization [15]. Limestone powder, class C fly ash, silica fume and water were used to manufacture bricks without cement [16]. The strength was observed to increase with increase in silica fume. At 20% silica fume, the strength was maximum at early curing days and increased from 23 to 26.5 N/mm^2 after 28 and 90 days of curing respectively. The cost of producing these bricks reduces by 6.5% when compared to those made from normal clay.

Another important way of increasing the strength of clay bricks is by firing them at a high temperature of between 500 and 1500 °C. Karaman *et al.* [17] observed that the time taken for firing and the firing temperature significantly affect mechanical properties of the bricks. The authors concluded that high temperature resulted in higher compressive and bending strengths as well as higher densities and lower water absorption. However, the duration of firing has relatively minimal effect on the mechanical properties of fired bricks. The compressive strength was observed to increase from 9.12 N/mm^2 at 700 °C to 31.4 N/mm^2 at 1100 °C. Laboratory study of clay bricks, prepared admixing cocoanut shell ash was carried out by Fernando [18], to determine the influence of the ash on strength of the bricks. The optimal Unconfined Compressive Strength (UCS) of 19.5 N/mm^2 was obtained at 2% cocoanut shell ash. AbdulKadir *et al.* [19] work on effect of coffee waste ash on the strength of fired clay bricks, and observed the compressive strength to decrease from 21.75 N/mm^2 for the normal burnt clay bricks to minimum of 10.24 N/mm^2 when mixed with 5% the ash. Tse [20] evaluated the suitability of clay deposits on the plain of Benue State for bricks production. The results showed that water absorption was between 3–4%, while the compressive strength ranges between 38 and 111 N/mm^2. The use of sawdust in production of light-weight clay bricks was also studied by Chemani and Chemani [21]. These bricks were fired to between 800 and 950 °C. The authors observed 9% optimum sawdust to give characteristics for clay bricks that are suitable for technological purposes.

There has been increase in the usage of clay bricks for construct of bungalows and other low-rise buildings for low to medium income earners in Nigeria, in the recent times. This has tremendously reduced the cost of buildings. However, cement is still used for construction of some elements of these buildings, such as reinforced concrete columns, slabs and beams. The use of clay pots in construction of slabs has reduced the use of cement in that regard. But columns and beams require materials with high flexural strength which is relatively minimal in normal clay bricks masonry. Although, even some traditional clay bricks making molds are made in such a way that the resulting bricks [22] have some kind of grooves and protrusions, the amount of flexural stresses building elements, made with them resist is very minimal. Attempt has also been made, although with little success, to produce unfired clay bricks incorporating some additives [23] that will add to their resistance to flexural stress. This study therefore, investigated the response of building elements constructed with fired clay bricks, made with special grooves and protrusions, which aided their assemblage in formation of the elements, to flexural stresses. This study indicated the possibility of using these bricks for short beams and columns in low rise buildings.

2 Materials and Methods

The materials used in this study are included specialized fabricated clay brick molds, whose assemblage consisted of protrusion and grooves (depression) with dimensions as shown on Table 1. Three categories of the molds, A, B and C were used, each with different dimensions as shown in Fig. 1.

Table 1. Brick dimensions

Brick no.	Groove (depression) size (mm)	Protrusion size (mm)
A1	100 × 55 × 110	100 × 53 × 108
A2	100 × 55 × 110	100 × 50 × 105
A3	100 × 55 × 110	100 × 47 × 102
B1	75 × 45 × 90	75 × 43 × 88
B2	75 × 45 × 90	75 × 40 × 85
B3	75 × 45 × 90	75 × 37 × 82
C1	60 × 35 × 70	60 × 33 × 68
C2	60 × 35 × 70	60 × 30 × 65
C3	60 × 35 × 70	60 × 27 × 62

The second major material used in the study was a clay soil, collected at Paggo village, along Minna-Paiko road in Niger state, Nigeria. The clay soil was transported to Civil Engineering Laboratory, Federal University of Technology, Minna, Nigeria, and prepared using the methods highlighted in BS 1377-1 [24].

The methods adopted in the study involved carrying out index properties, X-ray Fluorescence (XRF) test and X-ray Diffraction (XRD) tests, to classify the clay and obtain its mineralogical and oxide compositions. 15% water by dry weight the clay was then added and mixed properly and allowed for some time to give a good and uniform consistency. The mixture was then placed into the protruded and depressed molds and allowed to dry before removing them from the molds. The freshly demolded air-dried clay bricks are shown in Fig. 2.

The bricks were then transferred to Urban Shelter Company at Poggo village along Minna-Paiko road, where they were fired at 600 °C using industrial furnace, after which the bricks were brought out from their furnace. After firing, the color of the bricks changed to reddish (Fig. 3).

The fired bricks were then joined such that the protruded parts were fixed into the groves by cement slurry of 0.7 water-cement ratio. This fixture resulted to a clay bricks of between 0.60 to 0.66 m lengths as shown in Fig. 4.

The interlocked bricks were then transported to National Center for Agricultural Mechanization (NCAM) in Ilorin, Kwara State, Nigeria for compressive and flexural tests as shown in Figs. 5 and 6 respectively. Three samples each for the three categories were tested for both flexural and compressive strengths.

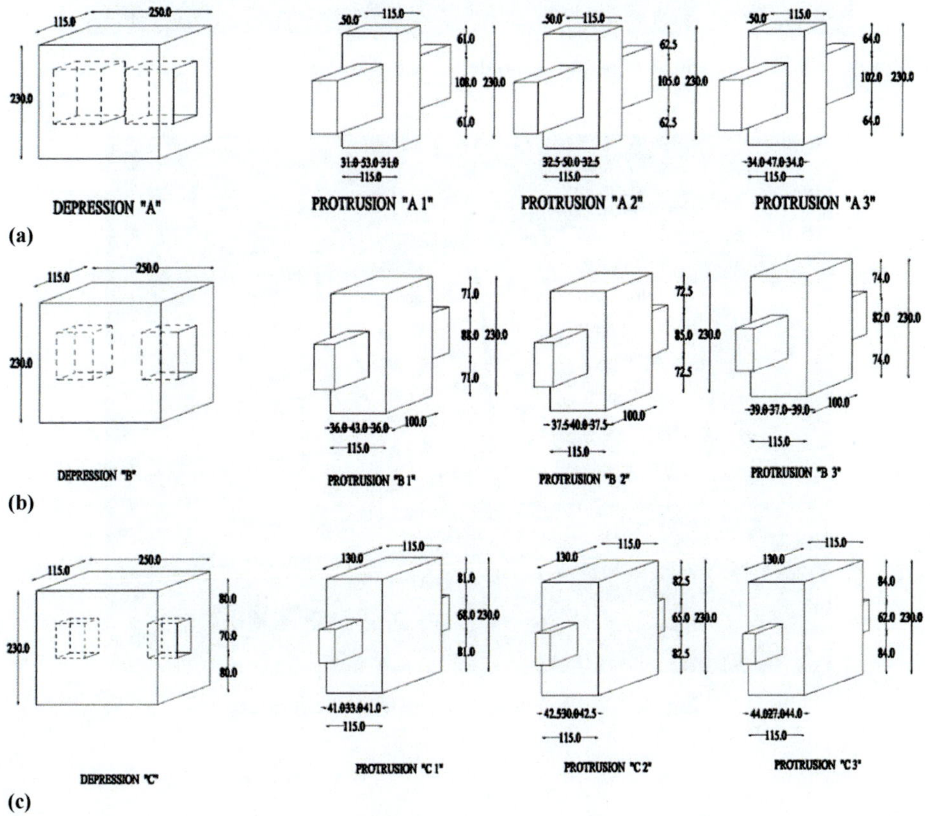

Fig. 1. Shapes and dimensions of the molds for the a: A category, b: B category and c: C category

Fig. 2. Freshly demolded clay brick components

Fig. 3. Clay bricks after firing from the furnace

Fig. 4. Interlocked clay bricks

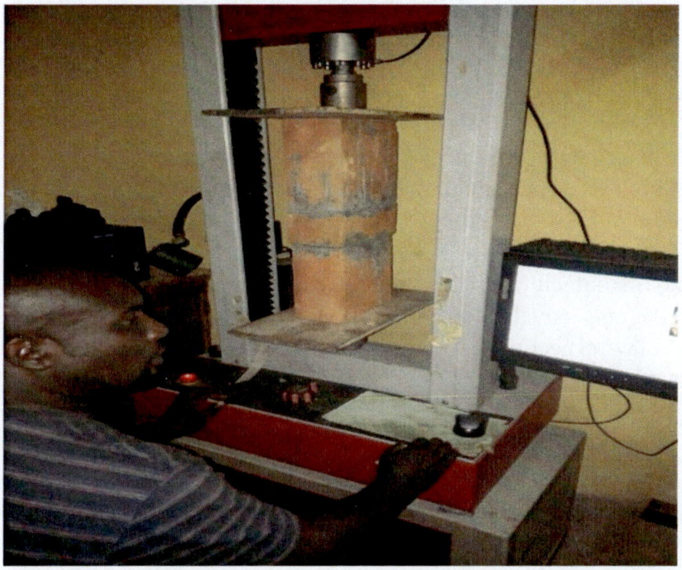

Fig. 5. Compressive test on clay brick

Fig. 6. Flexural test on clay brick

3 Results and Discussions

3.1 Index Properties

Results of index properties of the clay soil are shown on Table 2. From the results, the clay soil classified as clay of low plasticity (CL) according to Unified Soil Classification System (USCS) and A-6 according to AASHTO soil classification system. The liquid limit of 40.4% is slightly above the range of 25–38% recommended for clays to be used for clay bricks manufacturing. However, the sand, silt and clay compositions of 41.2, 39.4 and 19.4% respectively, are within the ranges of 20–45, 25–45 and 20–35% for sand, silt and clay respectively. The plasticity index of 10.3%, obtained for the clay is within the range of 7–16% recommended by Suryakanta [7].

3.2 Mineralogical and Oxide Composition of the Clay

The result of x-ray diffraction test conducted on the clay is shown on Fig. 7. From the figure, the clay has predominantly quartz, albite, montmorillonite and potassium mica. The presence of montmorillonite is probably responsible for the high liquid limit observed. Result of the x-ray fluorescence conducted on the clay is shown on Table 3. The clay contained substantial amount of silica and alumina with minimal traces of heavy metals.

Table 2. Index properties of clay soil

Description	Value
Natural moisture content	11.5%
Liquid limit	40.4%
Plastic limit	30.1%
Plasticity index	10.3%
Sand	41.2%
Silt	38.4%
Clay	20.4%
Specific gravity	2.73
USCS classification	CL
AASHTO classification	A–6

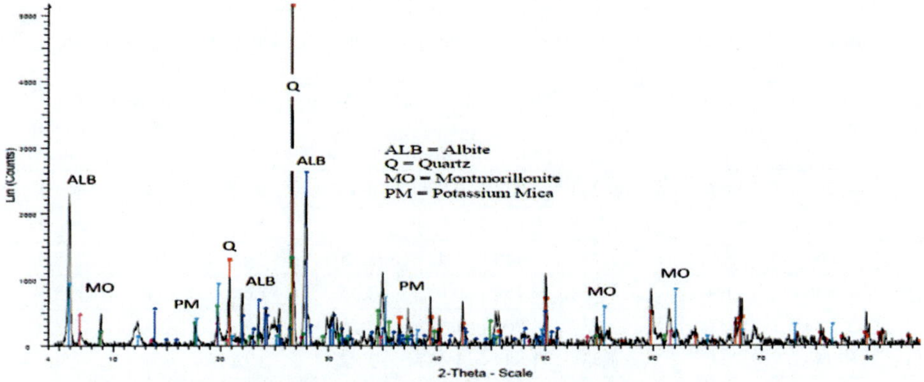

Fig. 7. X-ray diffraction graph of the clay soil

Table 3. Oxide composition of the clay soil

Oxide	Fe_2O_3	MnO	Cr_2O_3	V_2O_5	TiO_2	CaO	K_2O	P_2O_5	SiO_2	Al_2O_3	MgO	Na_2O	LOI	Total
Amount (%)	6.74	0.12	0.01	0.02	0.80	2.79	1.63	0.12	59.70	18.26	1.66	1.88	5.49	99.2

3.3 Compressive Strength of the Interlocking Clay Bricks

A typical result of the compressive strength of the interlocking clay bricks for category A is given on Tables 4 and 5 and Fig. 8.

Table 4. Compressive strength test result for category A clay bricks

Test no.	Date of test	Code	refz	Energy to peak (N/mm^2)	Energy to yield (N/mm^2)	Force at peak (N/mm^2)	Def. at peak (N/mm^2)	Force at yield (mm)	Def. at yield (mm)
1	06/11/2019	A1		111.5	111.5	65266.0	4.04	65260.0	4.04
2	06/11/2019	A2		174.0	131.0	79730.0	5.19	74460.0	4.64
3	06/11/2019	A3		103.7	103.7	64730.0	7.89	64730.0	7.89
Min.				103.9	103.9	84730.0	4.04	84730.0	4.04
Mean				129.7	115.6	89906.0	5.71	88150.0	5.52
Max.				173.9	151.0	79730.0	7.89	74460.0	7.89
SD				38.5	14.1	8511.4	1.97	5471.0	2.07
C of V				29.7	12.2	12.2	34.81	8.03	37.51
ICI				34.0	80.4	48763.0	0.80	54559.0	0.38
UGL				225.5	150.4	91050.3	10.61	81741.0	10.66

Table 5. Continuation of compressive strength test result for category A clay bricks

Test no.	Force at break	Deformation at break	Strain at peak	Strain at yield	Strain at break (N/mm^2)	Stress at yield (N/mm^2)	Stress at break (N/mm^2)	Stress at peak (N/mm^2)	Young's modulus (mm)
1	42250.0	4.60	0.734	0.734	0.837	2.47	1.60	2.47	363.5
2	67440.0	5.66	0.944	0.842	1.034	2.82	2.65	3.01	399.9
3	64260.0	7.91	1.434	1.434	1.437	2.45	2.43	2.45	235.3
Min.	42250.0	4.60	0.734	0.734	0.837	2.45	1.60	2.45	235.3
Mean	57983.3	6.07	1.037	1.004	1.103	2.58	2.55	2.64	333.0
Max.	67440.0	7.91	1.434	1.434	1.437	2.82	2.19	3.01	399.9
SD	13717.9	1.68	0.359	0.376	0.306	0.21	0.52	0.32	86.5
C of V	23.66	27.75	34.608	37.507	27.748	8.03	23.66	12.18	26.0
ICI	23905.7	1.58	0.148	0.059	0.343	2.06	0.90	1.84	118.1
UGL	92060.9	10.24	1.929	1.939	1.863	3.09	3.48	3.44	547.8

The graph of stress-strain curves shows that the stiffness of the bricks is low with gentle increase to a yield point and then finally to a braking point. This is a representation of the brittle nature of the fired clay bricks. The compressive strength of the three categories of the interlocking clay bricks is shown on Fig. 9.

The highest compressive strength of 3.7 and 3.4 N/mm^2 were recorded in category B for specimen B2 and B3 respectively. The size of protrution and depression in category A bricks are close, thus leaving little space to be filled with cement slury. In category C bricks, the thickness of the protrusions are relatively small which can easily give way under compression. Category B bricks however, posses medium thickness of protrusion

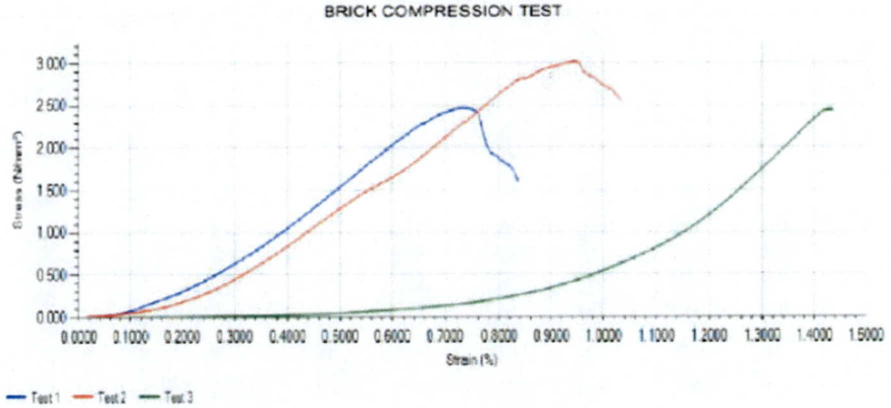

Fig. 8. Result of compression test for category A clay bricks

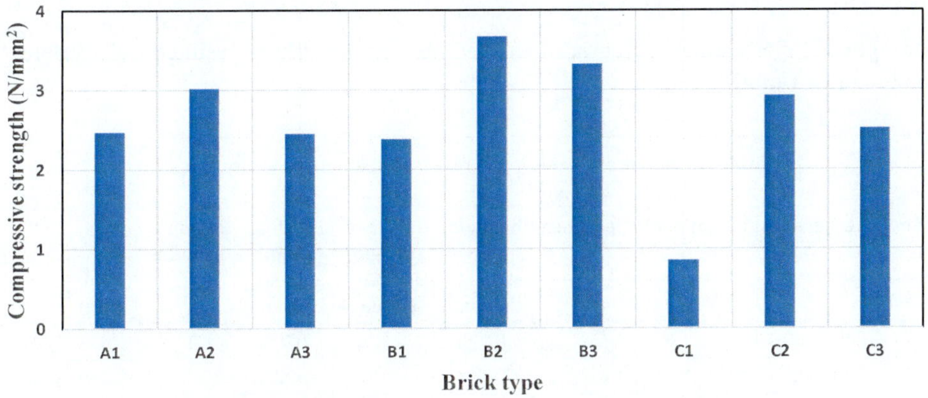

Fig. 9. Compressive strength of fired interlocking bricks

with B2 and B3 having higher spaces to be filled with cement slurry on interlocking. This is probably responsible for the higher compressive strength values recorded.

3.4 Flexural Strength of the Interlocking Clay Bricks

Result of flexural strength of the three categories of the interlocking clay bricks is shown in Fig. 10, while typical result as obtained from the compression machine is shown on Tables 5 and 6 and Fig. 11 for category A.

The result showed that flexural strength depends mainly on depth of the protrusions and to a large extent, on the space to be filled with cement slurry. The highest depth of protrusion (102 to 108 mm) exists in category A bricks, which is probably responsible for the higher flexural strength recorded for this category. Also, the higher spaces available to be filled with cement slurry occurs in A2 and A3 of this category which must have contributed to the higher strength (0.36 and 0.48 N/mm^2) recorded. The lowest flexural

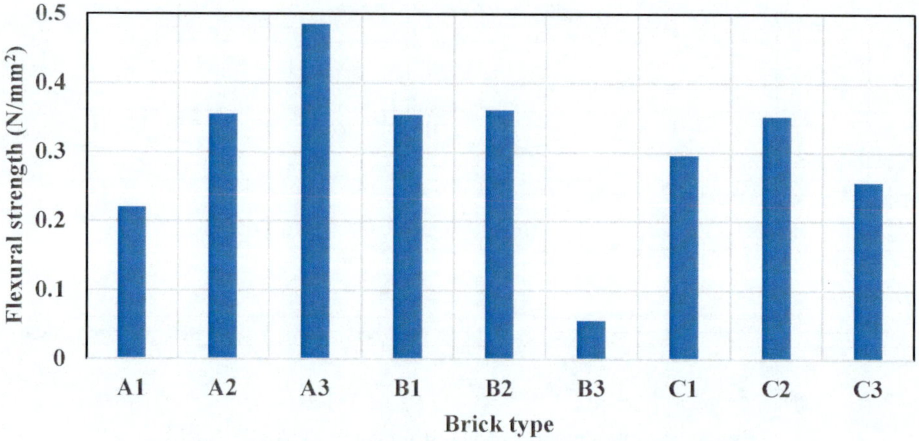

Fig. 10. Flexural strength for the three categories of the interlocking bricks

strength of 0.06 N/mm^2, observed with specimen B3 must have resulted from a defect on the brick (Table 7).

Table 6. Flexural strength test result for category A clay bricks

Test no.	Date of test	Treatment	Specimen	Bending strength at yield (N/mm^2)	Bending strength at peak (N/mm^2)	Bending strength at break (N/mm^2)	Bending modulus (N/mm^2)	Def. at yield (mm)	Def. at peak (mm)
1	06/11/2019		A1	0.220	0.220	0.220	41.3	0.973	0.973
2	06/11/2019		A2	0.184	0.184	0.097	44.0	0.869	0.869
Min.				0.184	0.184	0.097	41.3	0.869	0.869
Mean				0.202	0.202	0.158	42.7	0.921	0.921
Max.				0.220	0.220	0.220	44.0	0.973	0.973
SD				0.025	0.025	0.087	1.93	0.074	0.074
C of V				12.006	12.606	65.071	4.52	7.989	7.989
IGL				−0.027	−0.027	−0.625	25.4	0.260	0.260
UCL				0.431	0.431	0.942	60.0	1.582	1.582

From Fig. 11, specimen A1 and A2 showed similarity in slope, strain at failure and mode of failure. Specimen A3 however, showed lower slope with higher flexural strength and strain at failure. The higher strain at failure and higher magnitude of flexural strength is attributed to size of the protrusion and higher space available to be filled by cement slurry.

Table 7. Continuation of flexural strength test result for category A clay bricks

Test no.	Def. at break (mm)	Force at yield (N)	Force at peak (N)	Force at break (N)	Plastic strain at break (%)	Strain at yield (%)	Strain at break (%)	Strain at peak (%)	Transvers rupture strength (N/mm^2)
1	0.973	1784.0	1784.0	1784.0	0.005	0.537	0.537	0.537	0.220
2	1.238	1492.0	1492.0	784.0	0.464	0.480	0.683	0.480	0.184
Min.	0.973	1492.0	1492.0	784.0	0.665	0.480	0.537	0.480	0.184
Mean	1.105	1638.0	1638.0	1284.0	0.234	0.508	0.610	0.508	0.202
Max.	1.238	1784.0	1784.0	1784.0	0.464	0.537	0.683	0.537	0.220
SD	0.187	206.5	206.5	707.1	0.325	0.041	0.103	0.041	0.025
C of V	16.050	12.8	12.8	55.1	138.6	7.969	16.950	7.989	12.605
IGL	−6.578	−217.1	−217.1	−5069.0	−2.683	0.143	−0.319	0.143	−0.027
UCL	2.789	3493.1	3493.1	7637.1	3.151	0.879	1.539	0.873	0.431

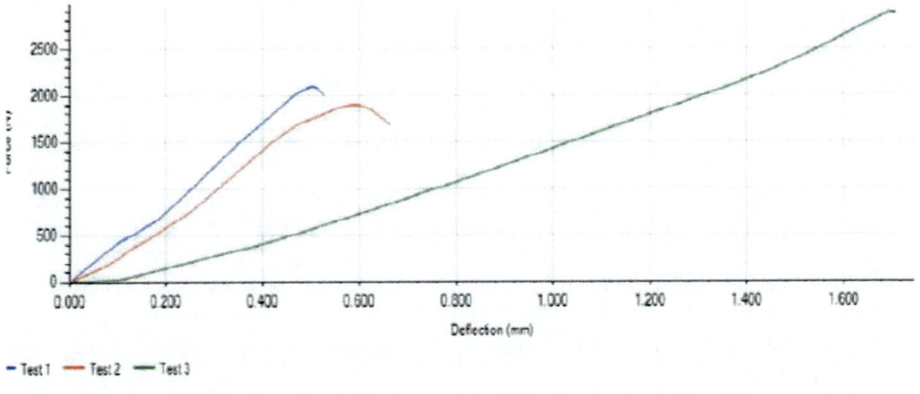

Fig. 11. Typical result of Flexural strength for category A interlocking Brick

4 Conclusion

The following conclusion is drawn from the study:

Elements formed, interlocking clay bricks of category B, which has higher space between the grooves and protrusions, for cement slurry binder, gave highest compressive strength of 3.7 N/mm^2, which satisfy the strength for load bearing walls according to Nigerian Industrial Standard.

Elements formed, interlocking clay bricks of category A, which has longer and deeper protrusions and grooves respectively, gave the highest flexural strength of 0.48 N/mm^2, which is far above the near zero flexural strength attributed to masonry clay bricks.

References

1. Ogunbiyi, M.A., Samson, R.A., Oginni, F.A.: Comparative study of cement stabilized clay brick and sandcrete block as a building component. Int. J. Appl. Sci. Technol. **4**(6), 56–61 (2014)
2. Omotoso, O., Simeon-Oke, O.O.: Escalating cost of cement a threat to a sustainable housing development in Nigeria. J. Res. Int. Bus. Manage. **2**(5), 97–101 (2012)
3. Priyanka, S., Devi, R., Raghava, E.V.: Partial replacement of cement with bentonite clay in concrete. J. NX-A Multi. Peer Rev. **4**(8), 12–14 (2018)
4. Nithiya, R., Chris, A.L., Vinodh, K.R., Anbalagan, C.: Experimental investigation on bricks by using various waste materials. Int. J. Latest Trends Eng. Technol. **6**(3), 395–402 (2016)
5. Kiptum, C.K., Hatangimana, V., Niyonagira, D., Nyirahabimana, S.: Physical properties of clay and bricks in Nyagatare, Rwanda. IOSR J. Mech. Civ. Eng. (IOSR-JMCE) **11**(2), 97–100 (2014)
6. Njau, H.G., Park, E.: Compressive strength of unfired composite bricks made of same clay and natural fiber of Tanzania. Int. Res. J. Eng. Technol. **2**(9), 13–15 (2015)
7. Suryakanta, N.: Field Test of Soil to Determine Suitability of Soil for Brick Manufacturing, CIVIL BLOG. ORG (2014)
8. Murali, K., Sambath, K., Mohammed, S.H.: A review on clay and its engineering significance. Int. J. Sci. Res. Publ. **8**(2), 8–11 (2018)
9. Miqueleiz, L., et al.: Allumina filler waste as clay replacement material for unfired brick production. Eng. Geol. **163**, 68–74 (2013)
10. Venkatesan, A., Anand, G., Fernandez, A.G., Natarajan, V.V.T., Alex, A.: A comparative strength and water absorption test on brick made of wood ash, charcoal with clay bricks: a comparative study. Int. J. Sci. Technol. **3**(3), 81–86 (2015)
11. Kumar, E.R., Hooda, E.N.: An experimental study on properties of fly ash bricks. Int. J. Res. Aeronaut. Eng. **2**(9), 56–67 (2014)
12. Chokshi, D.S., Pitroda, J., Makwana, A.H.: A competitive assesment on fly ash bricks in central Gujarat region of India using chi-square test through SPSS software. Int. J. Eng. Sci. Res. Technol. **3**(5), 320–331 (2014)
13. Smeu, S., Gal, A., Badea, C.: Environmentally friendly building materials: unfired clay bricks. J. Environ. **3**(3), 47–50 (2014)
14. Ali, A.F., Guney, C.O., Ali, A.F., Mojtaba, S.B.: Fundamentals of soil stabilization. Int. J. Geo-Eng. 1–16 (2017)
15. Naeini, S.A., Naderina, B., Izadi, E.: Unconfined compressive strength of clayey soils stabilized with waterborne polymer. KSCE J. Civ. Eng. **16**(6), 943–949 (2012)
16. Turgut, P.: Manufacturing of building bricks without Portland cement. J. Clean. Prod. **37**, 361–367 (2012)
17. Karaman, S., Ersahin, S., Gunal, H.: Firing temperature and firing time influence on mechanical and physical properties of clay bricks. J. Sci. Ind. Res. **65**, 153–159 (2009)
18. Fernando, P.R.: Experimental investigation of the effect of fired clay brick on partial replacement of rice husk ash (RHA) with brick clay. Adv. Recycl. Waste Manage. **2**(1), 1–4 (2017)
19. Abdulkadir, A., Hinta, H., Sarani, N.A.: The utilization of coffee waste in to fired clay brick. ARPN J. Eng. Appl. Sci. **10**(15), 6289–6292 (2015)
20. Tse, A.C.: Suitability of flood plain deposits for the production of burnt bricks in Parts of Benue State Central Nigeria. Geosciences **2**(2), 1–6 (2012)
21. Chemmani, H., Chemmani, B.: Valorization of wood sawdust in making porous clay brick. Sci. Res. Essay **8**(15), 609–614 (2013)

22. Dalkılıç, N., Nabikoğlu, A.: Traditional manufacturing of clay brick used in the historical buildings of Diyarbakir (Turkey). Front. Archit. Res. **6**, 346–359 (2017)
23. Benkhadda, N., Khaldoun, A.: Effective unfired clay bricks with natural additives. Int. J. Adv. Sci. Technol. **29**(02), 254–260 (2020)
24. BS 1377: Methods of test for soils for civil engineering purposes. British Standard Institution, London (1992)

Check for updates

Sustainable Planning for Provision of Basic Infrastructural Facilities in Rural Areas - Majgaon Village a Case Study

Umesh L. Deshpande[1](✉), Mohammad Iliyas B. Bagwan[2], and Anand B. Tapase[3]

[1] Department of Applied Mechanics, Government College of Engineering, Karad, Maharashtra, India

[2] Construction Management, Department of Civil Engineering, Government College of Engineering, Karad, Maharashtra, India

[3] Rayat Shikshan Sanstha's, Karmaveer Bhaurao Patil College of Engineering, Satara, India

Abstract. The increasing demand for water in combination with frequent drought periods, even in areas that are traditionally rich in water resources, puts at risk the sustainability of current living standards. The development of human societies is heavily dependent upon the availability of water with suitable quality and in adequate quantities, for a variety of uses ranging from domestic to industrial supplies. In the present study, an attempt is made to comprehensively understand the standards, culture, and livelihood of the Majgaon village based on various data interpreted from a socio-economical cum technical survey. After the interpretation of survey data, it was observed that Majgaon village is dealing with problems associated with solid and liquid waste and sewage water treatment and its disposal. However, untreated or partially-treated sewage poses an acute water pollution problem that causes low water availability. Pollution by inappropriate management of industrial and household wastewater is one of the major environmental problems the researchers are dealing with today. To address the household wastewater, adoption of new technologies and waste minimization initiatives are being encouraged. A development plan based on the identified problems for the village is prepared to evolve at an appropriate solution with the help of available local resources. The work is decentralized and on priority, the problem of wastewater and sewage waste treatment was studied and. For the management of wastewater and sewage waste disposal, the study and analysis of different methods were carried out. An analytical study for the integrated management of wastewater and sewage waste disposal of Majgaon village is completed with cost and time analysis. The sequencing batch reactor method was selected for the village as per the need and a detailed project report based on the analysis and feasibility is prepared. For the solution to excessive charges of drinking water pumping system, solar water pumping system has been designed.

Keywords: Sustainable rural development · Wastewater · Sewage water

A. Tapase et al. (Eds.): GeoChina 2021, SUCI, pp. 132–146, 2021.
https://doi.org/10.1007/978-3-030-79644-0_11

1 Introduction

India in the twenty-first century is an interesting mosaic of new and old challenges and opportunities. Huge development disconnects such as inequity in health, education, incomes, and assets, opportunities for growth, and access to resources. 70% of the population in India lives in rural areas, engaged in an agrarian economy with agriculture and allied sectors employing 51% of the total workforce but accounting for only 17% of the gross domestic product (GDP). Without rural development, India cannot optimally realize its growth potential and claim its place in the world.

To develop the country, we have to start from the villages. So, it is necessary to find out solutions to developmental issues in the social field by using research and skilled manpower from these institutes and application of appropriate technology. Unnat Bharat Abhiyan (UBA) is aiming at the participation of higher educational institutions to work with the people of rural India in identifying development challenges and evolving appropriate solutions for accelerating sustainable growth.It is necessary to analyze various social and developmental issues in the state through research and to find out perfect solutions to them. Accordingly, to enhance the research among various Government agencies and educational institutes on serious social and economic problems, it is necessary to boost the participation of students from engineering and other vocational courses in various project-wise schemes.

Socio-economic survey tools are designed to collect information as a means of improving understanding of local resource management systems, resource use, and the relative importance of resources for households and villages. Surveys also provide information on interaction with the government decision-making systems and community perceptions of trends and priority issues. Knowledge about community-based institutions, which is also obtained, and their roles in the sustainable use and conservation of natural resources, helps to facilitate or reinforce a consensus on land tenure and rights for the region, now and in the future. A socio-economic survey is regarded as one of the most important sources of statistical data on household expenditure and income as well as other data on the status of housing, individual and household characteristics, and living conditions.

2 Socio-Economic Survey

2.1 Socio-Economic Survey of Majgaon Village

A Socio-Economic survey is carried from Majgaon village. The data related to the solid waste management system, sewage treatment system, water supply systems, sources of drinking water, availability of toilets is done by collecting data per the questioner prepared.

2.2 Interpretation of Socio-Economic Survey

It is learned from the survey that the waste collection is carried with tractor-trolley daily from the doorsteps of every house which is disposed into the pits as the waste collection treatment is not available as mentioned in Table 1. Sewage water is carried through over the ground gutters into the streams but no treatment plant is available in Table 2. Water supply is through wells and river and the drinking water is available through public and individual tap waters Table 3 and Table 4.

From the information given in Table 5 and statistical representation in Fig. 1, it is observed that 97% of the total households of the village have their toilet in their house and 3% of villagers use public toilets. Almost every household has its own toilets.

Table 1. Solid waste management system

Waste collection	Not available
Collection schedule	Not available
The vehicle used for the collection	Not available
Disposal of waste	Open grounds
Treatment of collected waste	Not available

Table 2. Sewage water treatment system

Type of Gutter		Existing Condition
Kaccha		Poor
Pakka	Overground	Good
Treatment Plant		No
Disposal		Directly into Streams

Table 3. Water supply system

Name	Doorstep
Water supply Well	TCL powder
River	TCL powder

Table 4. Source of drinking water

Source of drinking water	Households	Percentage
Public tap water	64	46
Individual tap water	75	54
Total	**139**	**100**

Table 5. Availability of toilet

Availability of toilet	Households	Percentage
Households with toilet	141	97
Household without toilet	5	3
Total	146	100

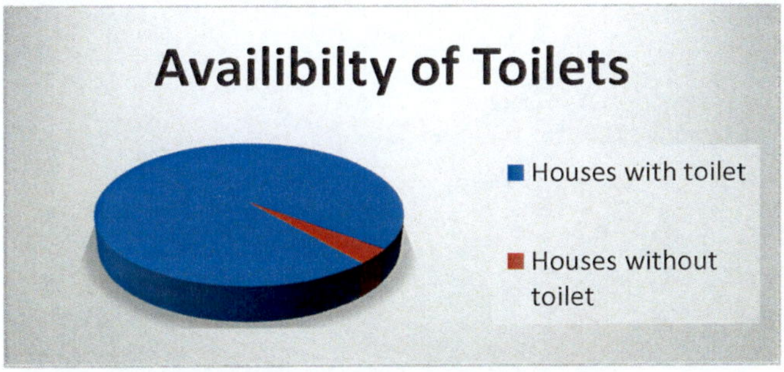

Fig. 1. Statistical representation of availability of toilets

Table 6. Renewable and non-renewable energy

Source	Households	Percentage
Renewable energy	118	42
Non-renewable energy	166	58
Total	284	100

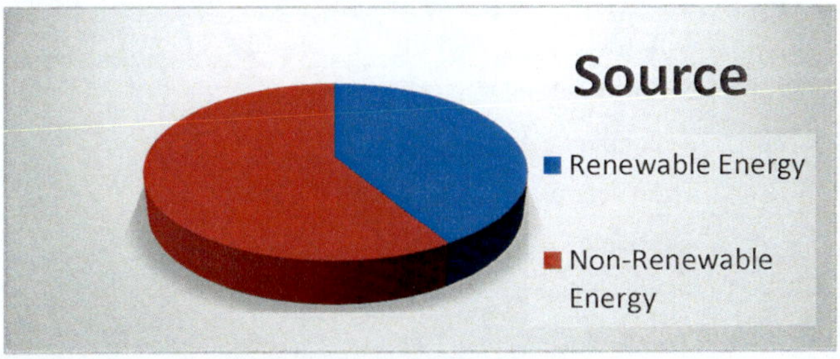

Fig. 2. Sources of energy

From the information shown in Table 6 and from Fig. 2, it is observed that 42% of the total households of the village use renewable energy for cooking and water heating and about 58% of households use non-renewable energy. Due to very less use of renewable energy, the consumption of non-renewable energy is high.

Table 7. Migration status of the village

Migration status	Population	Percentage
Temporary migrated	35	5
Permanent migrated	20	3
Native	685	92
Total	**740**	**100**

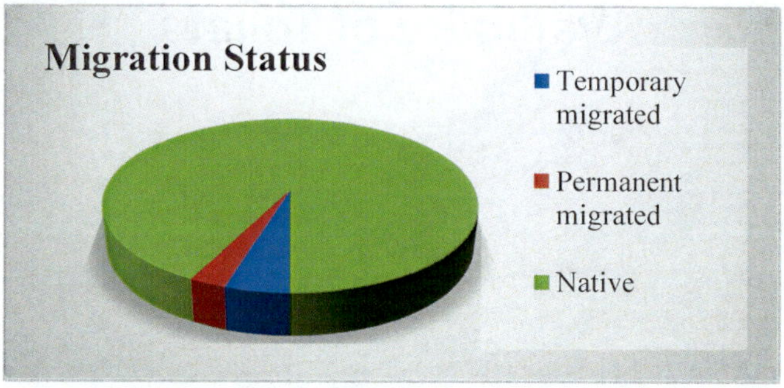

Fig. 3. Migration Status of the total population of the village

From the information shown in Table 7 and Fig. 3, it is observed that 10% of the total population of the village is migrated to different cities in search of a job and also for higher education.

Table 8. Literacy details of villagers

Literacy	Population	Percentage
Illiterate	114	15
Primary school	83	11
Secondary school	289	39
High school	137	19
Graduate	100	14
Postgraduate	17	2
Total	**740**	**100**

From Table 8 and Fig. 4, it is noticed that 15% of the total population of the village is illiterate. Those who are not able to write or read are considered illiterate. 39% have completed secondary education and only about 14% have completed their graduation. The percentage of Secondary education in the village is more and only about 2% have completed their post-graduation.

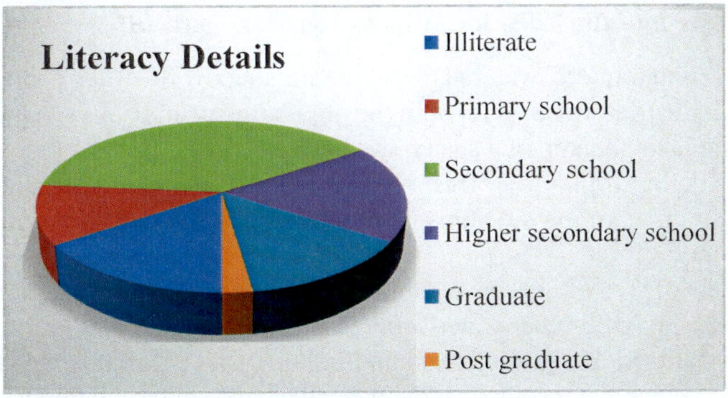

Fig. 4. Statistical representation of literacy details of villagers

2.3 Overall summary of Socio-Economic Survey

a) There is no treatment of solid waste and sewage waste. Sewage waste is directly disposed into streams; also, solid waste is dumped onto the open ground without any prior treatment.
b) Most of the houses are of load-bearing structures.
c) Since roof height is low, there is a lack of ventilation and natural light, which results in more demand for electricity for light and air conditioning.
d) The water supply pumping system of the village runs on MSEB electricity. The electricity cost is about 1 to 1.5 lakhs per annum.
e) All government buildings work on the electricity received from MSEB.
f) 6 houses in the village don't have a gas connection, which is about 4% of the total household.
g) A 24% population directly only depends on agriculture income.
h) 97% of total agricultural land is irrigated and the rest 3% of land un-irrigated. 89% of the land is irrigated by flooding methods.
i) The sex ratio of Majgaon Village is 980 women per 1000 men.
j) 85% of the population is literate and 15% is illiterate.
k) 8% of the population has been migrated to other cities.

3 Sewage Treatment Plant for Majgaon Village

3.1 Why India Needs Treatment of Wastewater

As India hurtles towards a more developed economy, one of the casualties has been the deteriorating state of our environment. Rapid industrialization has, unfortunately, hiked up the number of pollutants in our surroundings. One of these pollutants is wastewater. Management of wastewater through wastewater treatment plants in India has become an imperative of our cities today.

3.2 Wastewater and the Need for WastewaterManagement

Wastewater is contaminated water. The two main sources of water contamination are sewerage and industrial waste. With both the population of India and its industrial landscape increasing at a phenomenal speed, wastewater volume is also at an alarming rise. Adding to this is the shrinking of freshwater sources like rivers, wells, and groundwater and we have an alarming situation. Some fear that very soon water may become a premium commodity.

Rising levels of wastewater have another consequence. Not only it is unfit for consumption, it can mix with other water sources and contaminate it as well. For instance, contaminated water trickling into rivers and polluting it. When this water runs downstream and joins other water sources like other rivers, the contamination further spreads. Wastewater also seeps into the ground, contaminating underground water sources. The result is that almost every water source is today heavily polluted from rivers and wells to coastalareas.

3.3 Waste Water Management

With this situation, wastewater management is the best option. Since the process of reducing contaminants will takes decades of dedicated effort, the better solution would be to manage the wastewater. Fortunately, technology can help us here. Today we have sophisticated methods to treat wastewater. Here are some of the common methods used by wastewater treatment plants inIndia.

3.3.1 Physico-Chemical Treatment

Pollutants are usually classified according to size and different methods are used accordingly. For instance, larger particles are separated through gravity, flotation, or filtration. However, smaller particles are much more difficult to separate. This is where Physico-chemical treatment is particularly useful Chemicals known as flocculants and coagulants are used to separate them. This is a commonly used method to treat industrial waste. It is ideal for the removal of suspended matter like heavy metals, inorganic substances, oil, and grease, as well as dissolvedsubstances.

3.3.2 Biological Treatment

Unlike physio-chemical methods, the biological treatment uses organisms to treat pollutants. Over the years, scientists have developed various aerobic or anaerobic processes to treatwastewater.

3.3.3 Recycle and Reuse

In this method, wastewater is recycled using the membrane-based system. Membrane bioreactors use the simple science of ultrafiltration with a bioreactor to treat wastewater. In short, the method combines physic processes with biological methods. This is a commonly used method in industrial and municipal wastewater management. Treated water is recycled for various purposes, such as irrigation.

Zero Liquid Discharge System: Many people view this as cutting-edge technology in wastewater management. It removes all dissolved solids from the wastewater, giving us distilled water. Methods like RO are used to purify the water.

3.4 Types of Sewage Treatment Plants

There are many different kinds of sewage treatment plants that vary in the process by which they treat wastewater. Generally, they can be classified into the following types of system:

1. Activated sludge plant(ASP)
2. Rotating discsystem
3. Submerged aerated filter (SAF)
4. Tricklingfilter
5. Sequencing batch reactor(SBR)
6. Non-electricfilter
7. Suspended Media Filters(SMF)

3.5 Comparative Study of Sewage Treatment Plants

Different technologies for sewage treatment are in practice and varying claims on comparison of performance have been reported. Various studies have been carried out for the comparison of different sewage treatment technologies. Singh and John had carried out a comparison of Up-flow Anaerobic Sludge Blanket Reactor (UASBR) and SBR technologies for sewage treatment by studying different parameters which included BOD, COD, TSS, pH, and temperature. SBR showed better results than UASBR as per the treatment performance. But consideration of other factors such as power consumption and land requirement led authors to conclude that any of the two technologies can be used as per requirements and resource availability.

Life Cycle Cost (LCC) analysis has evolved as one of the measures for determining the suitability for use of a particular type of wastewater treatment technology. Different treatment processes can be used for any particular type of wastewater generated from any source. Most of the time the efficiency or the degree of treatment for a treatment technology may be almost similar to other technology or the difference may not be much significant. In such cases, Cost analysis may come of use to determine which treatment process can be preferred over others. Life Cycle Cost analysis helps to evaluate the cost of a treatment technology over its design period to help determine the most suitable one. This is particularly helpful in areas where the selection of wastewater treatment technology may be restricted due to financialconstraints.

Among the different new wastewater treatment technologies presently in use, UASBR has risen as most practical. In numerous nations, UASBR has been used for the treatment of high-quality wastewater, yet in India, it has been utilized for the treatment of residential wastewater. India is one of the main nations as far as the measure of sewage volume is treated by the UASBR process. It has been perceived as a standout amongst the most financially savvy and suitable sewage treatment procedures considering the natural necessities in India. At present, about 37 number of sewage treatment plants based on the UASBR are in operation inIndia.The present study was carried out to propose a rational basis for comparative evaluation of different sewage treatment processes based on different technologies. It was achieved by monitoring and comparing the performance of the STPs, evaluating and comparing the total cost involved in each of the treatment processes, and applying the LCC analysis technique to rationalize the comparative evaluation of STPs [9].

3.6 Results and Discussion

The performance of the Sewage Treatment Plants was evaluated over 4 months. The various parameters which were monitored included BOD, COD, Total Suspended Solids, Nitrates, and Coli form Reduction. The removal efficiencies concerning the mentioned parameters for each of the treatment plant were calculated and analyzed every week.

3.7 Life Cycle Cost Analysis

As mentioned, the performance evaluation is not alone sufficient for the comparison of different sewage treatment technologies. Life cycle cost (LCC) analysis has been largely applied.

as a tool to evaluate the best cost-effective alternative among various alternatives to achieve the lowest long-term cost of ownership. The net worth of investment cost analysis was done on actual existing plants based on the available information. This analysis was followed by LCC analysis to determine the life cycle cost for each of the STP based on the standard requirements of each technology. Table 9 presents the net worth investment cost for three technologies per MLD basis for business as usual conditions of the STPs.

Table 9. Net worth investment per MLD of the STPs [9]

Sr No	Parameter	Unit	SBR	UASB
1	Land area used	Acres/MLD	0.12	0.48
2	Capital cost	Crores/MLD	0.7	0.26
3	Biogas generation	m3/d	–	312
4	Annual power cost	Crores/MLD	0.0312	0.0052
5	Annual O&M cost (including recurring, chemical, manpower costs, etc.)	Crores/MLD	0.6	0.203
6	Total annual O&M cost	Crores/MLD	0.63	0.208
8	Cost of land	Crores/MLD	1.32	5.28
9	Unit capital cost including land	Crores/MLD	2.02	5.54
10	Annual interest	Percent	12	12
11	Economic life	Years	30	30
12	Capital Recovery Factor (CRF)		0.124	0.124
13	Total annual cost	Crores/MLD	0.88	0.894
14	Present discount factor		8.06	8.06
15	**Net present worth of investment cost**	**Crores/MLD**	**7.09**	**7.2**

It is observed that the present net worth investment cost of UASBR is the highest among both the reactors. The highest cost of UASB can be attributed to the fact that per MLD land provided in the present case was very high as compared to the standard requirement as prescribed by CPCB. The LCC analysis for both technologies per MLD basis is done as per standard area requirements as proposed by the CPCB using the formula explained above [9].

It is observed that based on prevalent conditions and actual area provided, the UASB based treatment plant shows the highest cost requirement per MLD as compared to SBR and MBBR based plants. It was observed that among UASB and SBR, the lowest cost requirement was shown by SBR, followed by UASB as per MLD of STPs.

In the present study also, when the standard conditions, especially the land area required for each of the processes, were used in the calculation, it was found that the SBR based plant showed the least cost required per MLD, which is in accordance with the prevalentstudies.

4 Solar Water Pumping System

4.1 Solar Water Pumping System for Majgaon Village

Panel Design for Majgaon Solar Water Pumping Systems
There are two open wells in Majgaon village under operation. These two wells are having a pump of capacity 5 HP. The total electricity bill for water pumping is about 1 to 1.5 lakhs per annum. Such extensive expenditure on electricity bills effects on minimizing the overall profit of Gram Panchayat. The solar water pumping system is one of the most suitable alternatives for the water pumping system. A solar water pumping system can be installed in the village. This will reduce the electricity bills as well as it is eco-friendly. The design of solar panels and stands over which panels are resting is as follows (Table 10).

Table 10. Estimation for the solar water pumping system

Total project cost	
A. Panel	= Rs. 115200
B. Inverter	= Rs. 80000
C. Wiring connections	= Rs. 50000
D. Stay frame	= Rs. 10798 × 3 = Rs. 32394
Subtotal	= Rs. 277594
Contingency 2%	= Rs. 5551.88
Total	**= Rs. 283145.88**

Solar pump system for 5 Hp motor requires 12 panels and the total estimated cost of the system is Rs. 283145.88.

4.2 Payback Period Calculations

The life of the plant will be 24 years.

Loan taken from bank = Rs. 2.84 lakhs. Interest on loan borrowed = 12% per annum. Electricity charges = Rs. 1 lakh per annum.

Installment amount = Rs. 1 lakh per annum including interest.

The payback period calculations are carried out here.

Table 11. Payback period for solar water pumping system.

Year	Amount	Interest Rate (12%)	Remaining Amount
1	Rs. 2,84,000	2,84,000 x 1.12 = Rs. 3,18,480	Rs. 2,18,480
2	Rs. 2,18,480	2,18,480 x 1.12 = Rs.2,44,697	Rs. 1,44,697
3	Rs. 1,44,697	1,44,697 x 1.12 = Rs.1,62,061	Rs. 62,061
4	Rs. 62,061	61,431 x 1.12 = Rs. 69,508	Rs. 30,491 (+)

From Table 11 and Table 12, it shows that the payback period will 3.5 years. Total recovery is of Rs. 3.18 lakhs, against loan of Rs. 2.84 lakhs.

4.3 Electricity Consumption

Table 12. Electricity consumption

Capacity of pump	= 5 HP
Conversion of HP to watt	= 5 × 746 = 3730 W
Daily requirement of electricity	= 3730 × 6 = 22,380 W per day
Yearly requirement of electricity	= 22,380 × 365 = 8.16 MW

The PV pumping system has many advantages that are summarized below:

(i) Low operating cost: No fuel required for the pump like electricity or diesel hence it reduces the operating cost.
(ii) Low maintenance: This technology requires Low-maintenance because of a lack of moving parts.
(iii) Harmonious with nature: Another important advantage is that it gives maximum water output when it is most needed i.e. in hot and dry months.
(iv) Flexibility: The panels need not be right beside the well. They can be anywhere up to 20 m away from the well, or anywhere you need the water. These pumps can also be turned on and off as per the requirement, provided the period between two operations is more than 30 s.
(v) Most important advantage is that technology is noiseless.

(vi) It is a Non-polluting technology, which means that it does not release greenhouse gases.

PV pumping system has its own limitations which are as following:

(i) Low yield: Solar pumping is not suitable for very high demand. It gives very low availability of maximum capacity 74.
(ii) Cost: The initial investment cost of this system is high.
(iii) Theft: Theft of solar panels can be a problem in some areas. So the farmers need to take necessary precautions.
(iv) Area/space: As the efficiency levels are low, the space required is relatively high.
(v) Solar energy is heavily dependent on atmospheric conditions.

5 Results and Discussion

5.1 Result and Discussion of Sewage Treatment Plant for Majgaon Village

Pollution by inappropriate management of wastewater is one of the major environmental problems in India. To address the household wastewater, adoption of new technologies and waste minimization initiatives are being encouraged.

The challenge thus is to find such low-cost, high-tech, user-friendly methods, which protect the degradation of our valuable natural resources. After a comparative study of wastewater treatment methods from the perspective of Life Cycle Cost and Time required for treatment of wastewater, the Sequencing Batch Reactor was found appropriate for Majgaon village.

5.2 Solar Water Pumping System for Majgaon village

The water pumping system of majgaon village is completely depending on electricity coming from Maharashtra State Electricity Board (MSEB). The expenditure on electricity is about Rs. 1 to 1.5 lakhs per year. Majgaon is a small village of Satara Dist. Spending around 1.5 lakhs on electricity only for pumping of drinking water is a big deal for such a small village. Hence, the installation of a solar water pump can reduce up to 55% expenditure on electricity for water pumping. The estimated cost for the overall solar water pumping system is Rs. 283145.88.

6 Conclusions

The Socio-Economical and Technical Survey form was developed for an effective investigation of various problems of Majgaon village. The survey form is processed and developed by referring to literature, field experience of various survey forms, and administrative experience to increase the participation of villagers for the overall upliftment of the village.

An attempt has been made to comprehensively understand the standards, culture, and livelihood of the Majgaon village based on various data interpreted from Socio-Economical cum Technical Survey Form.

Based on the observation of survey data following outcomes can be determined.

i. It was observed that the sewage waste generated is directly disposed to streams without any prior treatment which leads to acute water pollution problems that may cause significant health risks.
ii. It was observed that the domestic solid waste generated in the village is not collected at the village level and is directly dumped at a common point. No further treatment of dumped solid waste is carried out which results in biodegradation of organic matter. This will cause obnoxious odor and attract flies which are the chief agents of diseases like typhoid, diphtheria, diarrhea, etc.
iii. Water is supplied to villagers by adding TCL as a primary treatment through the pipeline network.
iv. The water supply system of the village works on MSEB electricity. The expenditure on electricity costs about 1 to 1.5 lakhs per annum.

From the survey conducted in Majgaon village, it was observed that there is no treatment of wastewater and it is directly disposed to streams. Because of the disposal of wastewater directly into streams, it pollutes the river water. This can affect the health of habitats severely. So, the study was carried out to provide an appropriate solution for the treatment of sewage waste for Majgaon village.

The challenge thus is to find such low-cost, high-tech, user-friendly methods, which protect the degradation of our valuable natural resources. After a comparative study of wastewater treatment methods from the perspective of Life Cycle Cost and Time required for treatment of wastewater, the Sequencing Batch Reactor was found appropriate for Majgaon village. The use of a Sequencing Batch Reactor is now being recognized as an efficient technology for wastewater treatment. Compared to the conventional treatment systems, SBR needs lesser material and energy, is easily operated, has no sludge disposal problems, and can be maintained by untrained personnel. Further, these systems have lower construction, maintenance, and operation costs.

Hence, designs with an estimation of Sequencing Batch Reactor for Majgaon village were prepared. Expense on electricity bills in the water supply system was high. So, the study was conducted for finding out an alternative solution for the water supply system. The solar water pumping system was found to be the best alternative for the water supply system. Hence, the design of a solar water pumping system was prepared. The proposal for the installation of a solar water pumping system was submitted to the Maharashtra Energy Development Agency (MEDA). Further MEDA has accepted the proposal and the tendering process is in progress.

Recommendations

1. Need to establish a wastewater treatment plant in Majgaon village.
2. Necessary to install solar or wind energy systems on all government buildings in the village.
3. For better transport, Panand roads should convert to rigid pavements.
4. Motivate and help villagers to use renewable energy.
5. Necessary to conduct educational workshops for modern agriculture, for better yield in less workforce and expenses.

6. Need to give subsidies for installing drip and sprinkler system for irrigation purpose, helps to minimize loss of water in farms
7. Necessary to make awareness about girl childbirth, to raise the sex ratio.
8. The requirement of conducting Praudh Shiksha Abhiyan.
9. Should provide the gas connection for those who rely only on the wood, dung cakes, agro residue, etc. for cooking.
10. The casing should be provided for the dinking pipeline to prevent contamination of potable water.

References

1. Eatwell, J., Milgate, M., Newman, P., Norton: Social Economics: The New Palgrave
2. Bhola, P.: Importance of socio-economic survey. JRPS Int. J. Res. Publ. Seminar **05**(02) (2014)
3. Iraq Household Socio-Economic Survey Manual, Organization for Statistics and Information Technology (COSIT) - Ministry of Planning, Government of Iraq (2012)
4. Government of India: National Sample Survey Office, Socio-Economic Survey
5. Sharma, R., Agrawal, P.: A case study on sewage treatment plant (STP), Delawas, Jaipur. Int. J. Eng. Sci. Comput. **7**(5) (2017)
6. Kulkarni, B., Wanjule, R.V., Shinde, H.H.: Study on sewage quality from sewage treatment plant at Vashi, Navi Mumbai. ScienceDirect Materials Today: Proceedings 5, pp. 1859–1863 (2018)
7. Hangargekar, P.A., Takpere, K.P.: A case study on waste water treatment plant, CETP (Common Effluent Treatment Plant). Int. Sci. J. **1** (2014). ISSN 2348-604X (Print) 2348-6058 (Online)
8. Kaur, R., Wani, S.P., Singh, A.K., Lal, K.: Wastewater production, treatment, and use in India. Int. J. Innov. Res. Adv. Eng. (IJIRAE) **2**(11) (2015)
9. Koul, A., John, S.: A life cycle cost approach for evaluation of sewage treatment plants. Int. J. Innov. Res. Adv. Eng. (IJIRAE) **2**(7) (2015). ISSN 2349-2163
10. Koul, A., John, S.: Comparative evaluation of sewage treatment plants: a novel approach. Int. J. Current Eng. Sci. Res. (IJCESR) **2**(7) (2015). ISSN (Print): 2393-8374, (Online): 2394-0697
11. Abdel_Kader, A.M.: comparison study between sequencing batch reactor and conventional activated sludge by using simulation mathematical model. In: Thirteenth International Water Technology Conference, IWTC 13, Hurghada, Egypt (2009)
12. Asiwal, R.S., Sar, S.K., Singh, S., Sahu, M.: Wastewater treatment by effluent treatment plants. SSRG Int. J. Civil Eng. (SSRG-IJCE) **3**(12) (2016)
13. Dhote, J., Ingole, S., Chavhan, A.: Review on wastewater treatment technologies. Int. J. Eng. Res. Technol. (IJERT) **1**(5) (2012). ISSN 2278-0181
14. Sundara Kumar, K., Sundara Kumar, P., Ratnakanth Babu, M.J.: Performance evaluation of waste water treatment plant. Int. J. Eng. Sci. Technol. **2**(12), 7785–7796 (2010)
15. Vrushali, S., Kaustav, C.: Sewage Treatment and Reuse - A Step Towards Water Conservation. Indian Council of Agricultural Research, New Delhi, India
16. Lu, S., Pei, L., Bai, X.: Study on the method of domestic wastewater treatment through new-type multi-layer artificial wetland
17. Mane, M., Patil, B., Pawar, M., Gohil, Y., Ghalimath, A.: Introduction to Waste Water Treatment by Root Zone Technique
18. Snehalatha, M., Busenna, P., Ratna Reddy, V., Anitha, V.: Rural Drinking Water Service Levels: A Study of Andhra Pradesh, South India

19. Rajagopalan, S.P., Narasimha Prasad, N.B.: Subsurface water in river beds as a source of rural water supply schemes
20. In association with the International Water Association (IWA) "Water resources and rural development"
21. Venot, J.-P., et al.: Farmers' adaptation and regional land-use changes in irrigation systems under fluctuating water supply, South India
22. Veerashekharappa, Bhide, S.: Sanitation strategies in Karnataka: a review
23. Wakode, P.N., Sayyad, S.U.: Performance evaluation of 25MLD sewage treatment plant (STP) at Kalyan. Am. J. Eng. Res. (AJER) **03**(03), 310–316. e-ISSN: 2320-0847 p-ISSN: 2320-0936
24. Khare, V., Nema, S., Baredar, P.: Status of solar & wind renewable energy in India. Renew Sustain. Energy Rev. **27**, 1–10 (2013)
25. Ananda, S., Rao, A.B.: Models for deployment of solar PV lighting applications in rural India. Energy Procedia **90**, 455–462 (2016)
26. Campana, P.E., Li, H., Zhang, J., Zhang, R., Liu, J., Yan, J.: Economic optimization of photovoltaic water pumping systems for irrigation. Energy Conv. Manage. **95**, 32–41 (2015)
27. Korpalea, V.S., Kolkata, D.H., Deshmukha, S.P.: Performance assessment of solar agricultural water pumping system. Int. J. Current Microbiol. Appl. Sci. **7**(05), 133–142 (2018). ISSN 2319-7706
28. Lu, S., Pei, L., Bai, X.: Study on PV water pumping for different application's. In: 2011 International Conference on Energy, Automation, and Signal (ICEAS), 28–30 December 2011. IEEE Xplore, 09 February 2012
29. Tiwari, A.K., Kalamkar, V.R.: Effects of total head and solar radiation on the performance of solar water pumping system. Renew. Energy **118,** 919e927 (2018)
30. Approach and Methodology for Socio-Economic Survey: Baseline and End-line. JRI Research, World Bank Document (2014)
31. Sindh Union Council and Community Economic Strengthening Support Program. EU-RSPN (2016)
32. New England Interstate Water Pollution Control Commission: Sequencing Batch Reactor Design and Operational Considerations
33. Imam, E.H., Elnakar, H.Y.: Design flow factors for sewerage systems in small arid communities. School of Sciences and Engineering, The American University in Cairo, Egypt. https://www.wateronline.com/doc/why-india-needs-more-wastewater-treatment-plants-0001

A Case Study of Slope Stability Assessment Thames River, London, Canada

Ron Xia[1](✉) and Hui Qian[2]

[1] McClymont & Rak Engineers, Inc., Maple, Vaughan, ON, Canada
[2] Chang'an University, Xi'an, China

Abstract. Three properties located on Adelaide Road, setting on top of the bank-slope of Thames River were selected as the project site for a case study. Bank slope failure caused by massive sliding and toe erosion was investigated. The mechanical regime of riverbank sliding, and toe erosion was studied. Methods of preventing slope erosion and massive sliding was discussed. Lessons learned from the case study was summarized. Following the guideline of OMNR and the regulations of long-term stable top of slope (LTSTOS) assessment of local conservation authorities is a right way to avoid an unexpected natural disaster to create a massive slope sliding.

1 Introduction

Three properties located at Adelaide Road (Highway 81), south east of Mt. Brydges, in the community of Delaware, City of London, Ontario, were selected as the project site for case study (Fig. 1). The three homes setting on top of a bluff, overlooking the Thames River, was constructed in 1969 (Brenk 2014). The bluff stands above an inward turn of a concave section of the tortuous meander's river - the Thames River. Depth from top of the bluff to the surface of river water was about 30 ft, measured on the day of June 11, 2015. Current of the Thames River is causing erosion of the bank, especially under flooding conditions during spring snow melt and rain rich seasons.

1.1 The Thames River

Thames River is one of the most southern Canadian water courses formed following the retreat of the Wisconsinan Glacier from Ontario and the upper reaches. The river is about 273 km long, as measured from the headwaters of the South Thames to Lake of St. Clair, and drains some 5,825 km^2 of land, making it the second largest watershed in southwestern Ontario (Nirupama and Simonovic 2006). Physio-graphically the Thames River basin divided into upper and lower portions at Delaware (UTRCA 1998). The land of the project site is within the boundary of Lower Thames Region Conservation Authority (LTRCA) (Brenk 2014).

A. Tapase et al. (Eds.): GeoChina 2021, SUCI, pp. 147–166, 2021.
https://doi.org/10.1007/978-3-030-79644-0_12

Fig. 1. The project site is located on Adelaide Road, south east of Mt. Brydges, in the community of Delaware, City of London, Ontario (From Google Map).

1.2 Geological Formation

The Lower Thames River basin has little relief, except for the incised Thames channel from Delaware to Thamesville. Much of the land surface is sand and clay plains and topography is flat to gently rolling. Two types geological plain formations dominated the lower Thames basin. The Bothwell and Caradoc sand plains are delta outwash deposits and the Ekfrid clay plain is lacustrine (MOE and MONR 1975).

The Upper Thames River basin is characterized by three lengthy moraines, numerous outwash deposits, drumlin fields and till plains. In general, the upper basin has more varied physiography with substantial reliefs. Reliefs are generally less than 50 ft and the moraines are discontinuous where they are dissected by drainages (MOE and MONR 1975).

Till plains with rolling topography constitute almost 60% of the land surface in this part of the basin. Drumlin fields are superimposed on the till plains. Moraines protrude through the till plains and form much of the western, southern, and southeastern basin boundaries.

Sand and gravel deposits in the London area were transported by melt waters and deposited as deltas where the spillways entered the glacial lake. The sand and gravel deposits are locally thick and extend to bedrock in the vicinity of the Fanshawe Dam.

Overburden covers all the lower basin and restricted bedrock outcrops occur only at Beachville and St. Mary's in the upper basin. The overburden thickness varies throughout the basin with a maximum reported thickness of approximately 300 ft in the moraine south of London (MOE and MONR 1975).

1.3 Soil Properties

Soils of the project site was developed in soil parent materials ranging in texture from heavy clays to coarse gravels. Many of the differences in texture and soil structure have been influenced by the various processes of material transport/deposit. A diagram showing grain size analysis for a sample taken from the site are shown in Fig. 2. Results indicate that the sample soil is composed of 0.3% of gravel, 41% of sand, 33% of silt and 25.7% of clay. Soil samples taken for laboratory analysis and testing were collected on June 8th/2019. Locations of soil samples taken refer to Fig. 5.

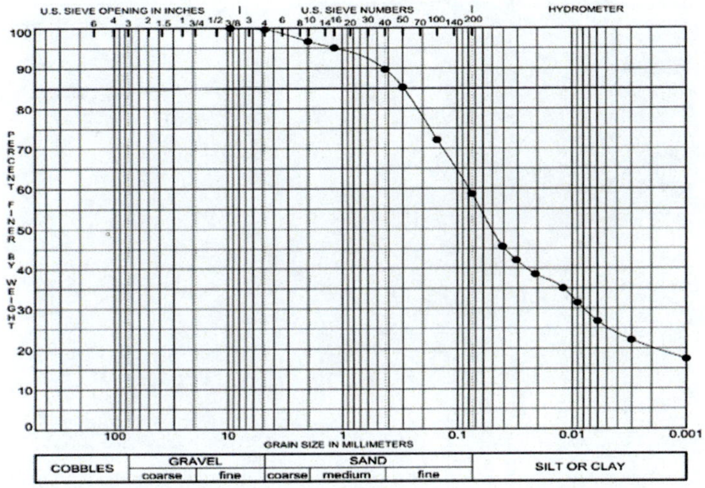

Fig. 2. Gradation curve of a soil sample taken from the site at the bottom of the riverbank within the zone of slope toe, by the author, on June 8th, 2019.

The original soil parent materials are highly calcareous and alkaline. However, the soils have developed on these materials are less calcareous as the leaching action of water on soil bases, especially calcium.

This leaching action, along with associated soil weathering, causes the development of soil horizons near the soil surface. These horizons differ from each other in properties such as texture, colour, thickness, structure, and consistence (Hagerty and Kingston 1992).

Variations of freshwater infiltration and soil drainage conditions made difference in soil formation and development from same parent materials.

Following the leaching action, the salts that originally contributed to the bonding of the particles were slowly removed by freshwater filtering through the ground. This may create the possibility to form a weak but water-rich sediment in clayey soils. The increasing in pore-water pressure, especially during periods of high rainfall or rapid snowmelt period may stimulate the favour conditions of riverbank erosion. Soil movement and porewater pressure increasing may further trigger an unexpected slope sliding.

2 Slope Conditions

The bank slope of Thames River, where three homes setting on top with rear yard facing southwest, fronting the river valley, was selected as the study site. Location of the site is in Lower Thames River water basin and close to the boundary of the Upper Thames River in the community of Delaware.

The site is in a critical condition of falling into the Thames River, as erosion of the banks is creeping ominously close to the homes (Fig. 3).

Fig. 3. Three homes on Adelaide Road are close to falling into the Thames River (Picture taking on April 17, 2013 by Mike Hensen 2013).

Home at bottom right in Fig. 3, was owned by David and Susan Shuttleworth. They bought the property and moved into house in 2001. The house was built up in 1969, right after the permit was issued. Backyard overlooking the Thames River was 40 ft long, if measured from the house to the top line of the riverbank slope (Patis 2014).

2.1 Slope Collapses

Problems of channel erosion and stream bank stabilization of Thames River was noticed in early 1971 and well documented in a report of "Flood and Erosion Control Works on The Lower Thames River From Chatham to Delaware" prepared by James F. Maclaren Limited (TRIC 1975).

Toe erosion making slope from unstable to collapse usually goes unnoticed until a significant amount of soil mass from the bank slope has been removed. A severe slope collapse was happened on March 13 to 14, 2009. A large part of land slid into Thames River and left the remain of back yard about 20 ft long in between the house and the newly formed slope top line (Brenk 2014; News Local 2009).

Falling soil piled on the toe of slope was gradually taken by river flood as solute, suspension sediment and bed load flowing transported to downstream. Toe erosion always strongly related with channel flow direction and varied with current velocity which is a function of peak flow banned with river flooding.

Another three years after the notice of the slope erosion to the land where their home standing on, a massive sliding into Thames River happened again and the owner reluctantly moved out in 2012. The further erosion was continuing and the Shuttleworth's home, showing at right bottom side in Fig. 2, was demolished on February 17, 2015 (Brenk 2015).

2.2 Peak Flow and Flooding

Floods may occur at any time of the year. Spring floods, caused by a combination of ice, snowmelt, and rain, are most frequently. Summer floods caused by major thunderstorms were experienced in the past. Probability analysis indicated a severe flood induced by tropical hurricane is also a possible occurring in autumn.

Generally, the highest monthly stream flows occur in March, with a few stream's discharging peaks in April. Usually, much higher flows occur during the March and April freshets than in other months. Freshets are the sudden raises in water level of the river, or a flooding caused by heavy rains and/or the rapid melting of snow and ice. Typically, summer floods have sharper peaks than the more frequent spring runoff floods.

Floods are closely linking with local climate and weather conditions. The prevailing winds in the region are westerly. Normal annual precipitation in the basin generally increases from the lower reaches to the headwaters, ranging from 32 in. at Chatham to 39 in. at Stratford. There is notably little seasonal variation in precipitation, with the difference between maximum and minimum, normal monthly precipitation being generally less than one inch (MOE and MONR 1975).

The snow belt area of southern Ontario is centred north of the Thames watershed, however its effect extends into the basin, causing an annual average snowfall of approximately 40 in. in the western end of the basin and approximately 68 in. in the eastern end (MOE and MONR 1975).

Peak flow is closely related to precipitations during a year and can be explained by a series of historical observations. We adopt the historical hydrological process in 1997, recorded in two hydrologic observation stations at London area, as an example. Plotting the observed total precipitation (rainfall and snowfall) versus river discharge measurements to illustrate the actual trend of peak flows during the year of 1997 (Fig. 4).

Hydrologic observation station, Thorndale is located upstream of London and hydrologic observation station, Byron is located south of London at the outlet of Upper Thames River Watershed (UTRW).

The differences in peak flows between Thorndale and Byron, when plotted on Fig. 4, show a similar pattern over the year. Discharge response to total precipitation occurred during the entire year of 1997.

2.3 Groundwater and Seepage Flow

Once precipitation infiltrates to the water table, it becomes ground water. Below the water table all materials are saturated. Soil type, composition and grain size of the subsurface material determines the availability of a seepage flow to Thames River as a base flow. At the project site, groundwater mainly recharged by the inflow from nearby higher area and by precipitation.

Surficial sand and/or gravel deposits flank the Thames River over most of its course. Extensive sand and gravel deposits along the Thames River at the Delaware section,

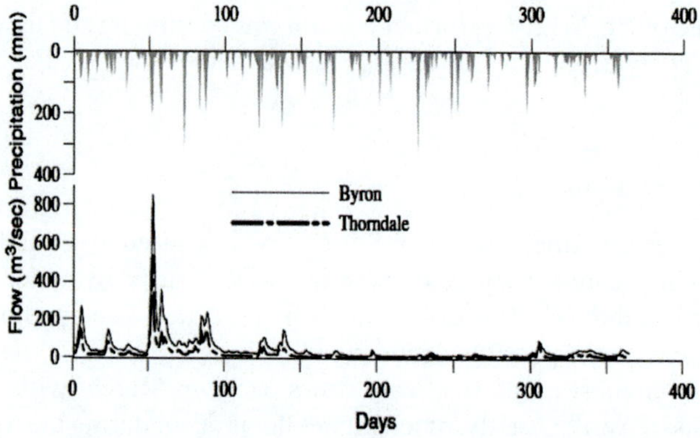

Fig. 4. Observed 1997 hydrographs at Byron and Thorndale and total precipitation at London, Ontario -- from Nirupama and Simonovic (2006).

Fig. 5. Groundwater seeping out from sandy soil was observed at project site (picture taken by Ron Xia on June 11, 2015).

especially at the area of our study, contributes groundwater discharging to the river. Phreatic surface was high and groundwater draining/seeping out at upper zone of the bank slope was observed during our site visit on June 11, 2015 (Fig. 5).

Groundwater levels reach a maximum in early spring, usually April, followed by a gradual decline to a minimum in late fall, usually October. It is uncommon for summer precipitation to result in a water level rise, although rates of decline may be slowed. This annual trend appears to have same pattern of peak flow in Thames River.

Annual groundwater fluctuation, in Thames River watershed, depends on precipitation, surficial soil type/composition, and the thickness of aquifer deposit in overburden. Shallow aquifers have annual fluctuations in the range of 3 to 10 ft. Annual water level fluctuation in deeper overburden aquifers are between 1 and 2 ft while water levels in deeper bedrock aquifers vary only a few inches (MOE and MONR 1975).

The exchange between groundwater and river water is relatively complex. Generally, river water is mainly recharged by groundwater at the study area. Water table/phreatic line, at the slope section, can be estimated by using Darcey's Law under fully saturated conditions (Fig. 6).

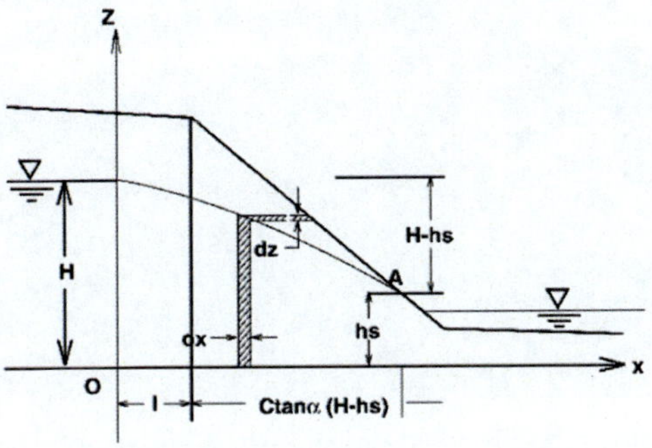

Fig. 6. Profile of a natural slope showing phreatic line computation and seepage zone (Xia et al. 2002).

Discharge per unit width along the slope face can be calculated as

$$q = -kz(dz/dx) \tag{1}$$

Where k is the hydraulic conductivity of the fully saturated zone, and dz / dx is the hydraulic gradient (Fig. 5). Solving Eq. (1) by integrating with boundary conditions of $x = 0, z = H$, we have:

$$q = k(H^2 - z^2)/2x \tag{2}$$

where H is the distance from baseline to water table determined during geotechnical investigation or measured from ground water observation system.

Equation (2) may settle the position of water table or phreatic line through a natural slope profile and can then be used to quantify the full saturated zone of seepage.

Direction of groundwater flow is generally from higher to lower elevations and toward the channel of Thames River in our study site. However, the direction of groundwater flow is influenced by subsurface structures and underground barriers. Influence of house basement and swimming pools on groundwater flow patterns was confirmed.

2.4 Subsurface Barrier

In common, the three properties, selected for our study of bank erosion and slope stability, all have one level of basement, septic system and private water wells. Wells were used for drinking water in the old days and for watering grass after tape water facility was installed. Two of the three properties have swimming pools in their back yard (Fig. 3).

Substructures of basement, swimming pool and septic tanks are barriers for groundwater seepage flow when water level is higher than the bottoms of the substructures.

Fig. 7. Piping and mudflow was observed on bank-slope surface of Thames River at the study site (picture taken by Ron Xia on June 11, 2015).

These subsurface barriers will alter seepage flow direction, rise groundwater level, change flow path and flow velocities.

Groundwater flow will slow down at the front of the substructures, raising water levels at local areas and seeking for new paths where soil is less competent to generate concentrated flow. The raised groundwater level and increased velocity at the area where saturated fine sand, silty sand, sandy silt and/or clayey silt is encountered, dilation/quick condition may happen at the surface of the bank slope, due to the increased hydrostatic pressure and hydraulic gradient (Fig. 7).

The impact of substructure on groundwater regime was found during our investigation. Caves formed by piping water at local area was noticed. Wetland plants reed-grasses growing on the bank slop top line were observed (Fig. 8). Wetland plants or vegetation cover the top of the riverbank slope would be an evidence of groundwater rising induced by subsurface burrier at the side of a former residence.

Increase in water content, softening clay bond or other cementing agents, decrease shearing resistance. Sheet erosion and rill erosion was detected at the areas of groundwater piping out (Figs. 5 and 7). Soil erosion stimulated the processing of riverbank sliding.

2.5 Freezing and Thawing

The climate of the Thames River basin is temperate, and the proximity of lakes Erie and Huron exert a moderating influence on temperature extremes in the basin. In general, temperatures decrease from the lower reaches at Lake St. Clair to the headwater area. The lowest mean monthly temperature occurs in January, ranging from about −3.9 °C at Chatham to −6.7 °C at Stratford, while the highest mean monthly temperature decreases from 22.2 °C at Chatham to 20 °C at Stratford (MOE and MONR 1975).

Fig. 8. Reed-grass grows on bank-slope top line of Thames River at the study site (picture taken by Tudor and Ron Xia on June 11, 2015).

Approximated estimation of air-freezing index at the area, project site is located, would be in the range from 450 to 500 degree-days Celsius. Air-freezing index is a summation of the daily mean degree-day for the freezing period. A long-term mean (30 year) air freezing index can be estimated from monthly mean air temperature data published by Environment Canada (CGS 2006).

Local long-term statistic weather conditions of the site observed at the nearby weather station in Delaware community are shown in Figs. 9 and 10. The 30 years' monthly-average weather conditions were recorded from 1982 to 2012, at the Strathroy-mullifarry weather station, Ontario, Canada (Statistics D. O. 2015).

Monthly average high temperature is 27.5 °C in July and monthly average low is − 8.1 °C in January. Three months from December to February monthly mean temperature is sub-zero. Ground freezing start in December and thawing start at early March. Three higher total monthly precipitations 111, 90 and 88 mm happened in June, December, and January – see Figs. 9 and 10.

During long period of wintertime ground is freezing. Segregation ice may form in frost susceptible soils, like organic mixed with sandy/clayey silt. Along with freezing front development frost heaving generated increasingly. Following ground is frozen, thermal creaks may happen on top layer of the riverbank slope, if a frozen soil shrinking stress exceeds the maximum tensile strength of the frozen ground (Xia 1985).

In raining seasons, water may easily infiltrate into the creaks formed in winter and generate a concentrated rill flow. The rapid water flow may cause internal/external soil erosion with steep hydraulic gradient to loosen soil material, to create an open cut channel and to make erosion area dramatically enlarge.

Freezing and thawing processes may supply stimulate conditions to initial soil erosion and develop a massive movement causing collapse of a natural slope.

In springtime, after snow disappeared, frozen ground thawing starts. Meltwater initially entering down below frozen soil (Woo and Marsh 1990) will refreeze to form

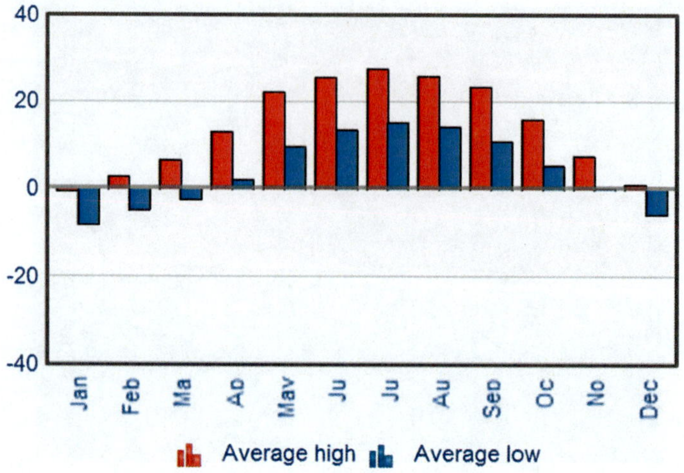

Fig. 9. Statistic overview of monthly average temperature at Delaware, Ontario, Canada (Statistics D. O. 2015)

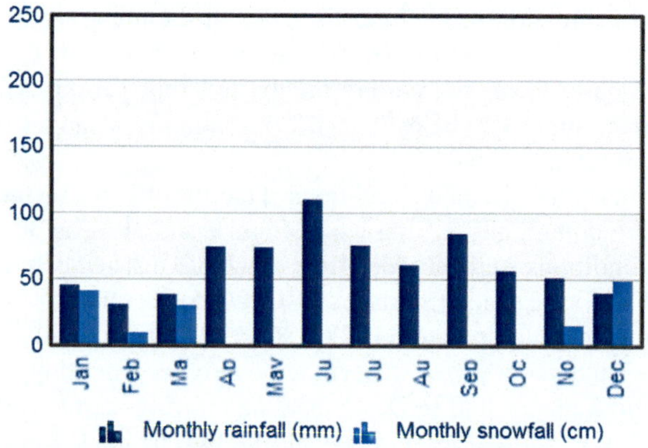

Fig. 10. Statistic overview of monthly average precipitation at Delaware, Ontario, Canada (Statistics D. O. 2015)

ground ice until the materials become impervious to further infiltration. During the re-freezing process, ice parcels and thick ground ice layer formed.

Thawing of the ground ice will release enormous voids, which will loosen soil structure, reduce shear strength and cohesion force, and the water released may act as a lubrication to reduce friction force or change friction angle at thawing front. This may strongly stimulate shallow slope failure as linear sliding and/or circular rotations.

Soil freezing depth ***X*** perpendicular to the surface of the bank slope, can be estimated with an extension to Stefan's approach and consider the release of latent heat associated with the formation of potential ice segregations:

$$X = \sqrt{\frac{2(k - SpL)Is}{Ls}} \tag{3}$$

where *Sp* is the value of the segregation potential, ***L*** is the volumetric latent heat of water to ice, *Ls* is the latent heat of soil, *k* is the thermal conductivity of the frozen soil, *Is* ground surface freezing Index and can be estimated from air-freezing index (CGS 2006). Thermal conductivity could be calculated by using the volumetric fractions of the various components of the soil (Farouki 1981):

$$k = \prod_{j}^{n} k_j^{f_J} \tag{4}$$

where *f* is the volumetric fractional content of soil component *j* which includes mineral materials (***m***), organic matter (***o***), water content (*w*), ice content (*I*) and air in voids (*a*). Their respective thermal conductivity (in W m^{-1} $^{o}C^{-1}$) are $k_m = 2.93$, $k_o = 0.25$, $k_w = 0.57$, $k_i = 2.20$ and $k_a = 0.025$. For a certain location of a slope the above components may vary with time and depth, depending on the hydrological and thermal conditions of the soils (Xia et al. 2001, Woo and Xia 1995a, b).

Coarse material with abundant interstitial voids provide ample storage capacity for the infiltrated rainwater and melt water, and their high hydraulic conductivity permits easy water movement into, out of, and within the coarse materials (Xia et al. 2002; Woo and Xia 1995a, b). Sources of seepage water may be from rainfall, snow melt or ground ice melt (Xia 1993) as well ground water from upper tableland.

Fine material may gradually leach out along with the fast flow speed of seepage and deposited at the lower portion of the slope foot or washed away by flowing water in the Thames River (Fig. 4). Velocity of groundwater seeping out can be described as:

$$V = K\frac{\Delta h}{\Delta L} \tag{5}$$

where $\boldsymbol{\Delta h}$ is difference of hydraulic energy potential, ΔL is length of flow path. Initial at the thawing season, $\Delta L = X$, which is the maximum frozen depth/thickness of the bank slope, if water seeping out the slope. Along with thawing depth growing the magnitude of $\Delta L = X$ is turning to smaller and smaller.

Velocity of seepage growing faster and faster. As ΔL changes from X to 0, the velocity V will turn to infinite. In reality, this velocity cannot be infinite but will take on extremely large values and cause fine material escaping.

Fine material escaping and pore water status changing may create a commendatory condition to form localized collapse and stimulate slope erosion. Seepage erosion is also named as subsurface erosion and was well documented (Xia et al. 2002) and widely observed at the surface of bank slope at our project site (Fig. 7).

2.6 River Water Under Cutting

As rivers and streams flow along seeking the path of least resistance, they develop different channel patterns in response to the physiography of the area. The Thames River takes on several channel patterns, but overall is best described as having a *sinuous channel pattern* or *irregular meanders (TRBSRT* 1998*)*. There are, however, stretches where the meandering is more regular such as near the site of our study (see Fig. 11) and reaches where the river is fairly straight, especially downstream of Chatham (TRBSRT 1998).

A variety of erosion and depositional forms can be found as bank/slope erosion, slumping, gullies, pools, and riffles at the study site. Bank erosion is visible at the outside meander where the three houses are located on top of the steep bluff (see Fig. 2).

The vertical movement of earth material downhill is termed mass wasting. It can either occur by slide (fast) as evidenced by steep-sided un-vegetated banks, or by creep (slow). Trees grow with curved trunks, adjusting their angle to the sun as they slip further down-slope.

Rotational slips were observed in the area where material from the top of the bank slides down leaving a concave shaped bank and undercut bluffs. Erosion rates are not steady and there may not be any change in a bank for many years until a single large discharge passes through and scours the bank with enormous energy (TRBSRT 1998).

Streams of the Thames River cut through the sandy soil down to the base level of the river, creating steep-sided narrow *gulches* along its lower course. *Pool and riffle* formations are twisted. This alternating of deep (pool) and shallow (riffle) areas tends to be continuing a relatively long-term process (Fig. 11).

Fig. 11. Aerial view of tortuous meanders at the section of project site of the Thames River (picture taken by Ron Xia on June 11, 2015).

3 Developable Limits

With the intensification and expansion of urbanization along valley lands and stream corridors, the risk of exposure to natural hazards has increased tremendously in the past few decades (Botzan 2011). Under the Conservation Authorities Act, a permit is required for waterway alteration, grade change or construction in natural hazard areas. Houses standing on top of river bank slopes to obtain a permit, geotechnical assessment to define the developable limits is required (Botzan 2011).

Of particular essence in determining the developable limits is the long-term stable top of slope (LTSTOS) line (Botzan 2011). LTSTOS is an imaginary projection, over a

100-year span, of the existing top of slope (ETOS). A guideline issued by the Ontario Ministry of Natural Resources (OMNR) is utilized to define the LTSTOS.

Three types of allowance of erosion hazard limit, described in the Guideline of Ontario Ministry of Natural Resources (OMNR), are of reliable for engineering planning, designing and interest to conservation authorities' review. They are simply summarized as 1) toe erosion allowance, 2) stable slope allowance, and 3) erosion access allowance (OMNR 2002). These three elements are schematically illustrated on Fig. 12.

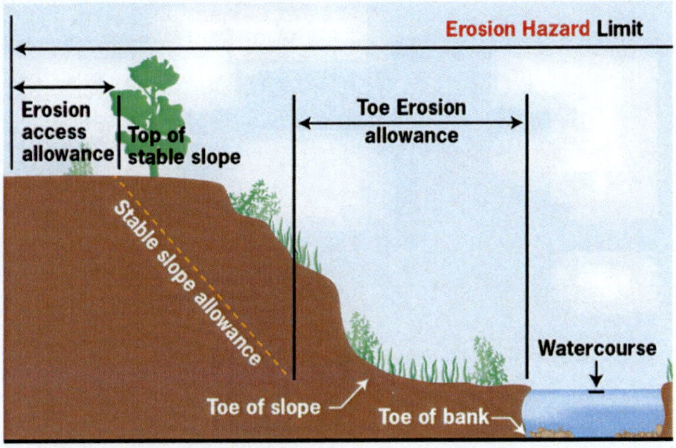

Fig. 12. Confined System, Erosion hazard limit where toe of valley slope is located less than 15 m from the watercourse (OMNR 2002).

3.1 Stable Slope Allowance

Geotechnical assessment of the LTSTOS is required in the light of the aforementioned OMNR Guideline (TRCA 2007). Slope stability verification for the section at study site was carried out using a professional commercial software "Slid-2018" developed by Rocscience Inc. for circular slip surfaces. A slope with a Factor of Safety (F.S.) equal to/greater than 1.5 is considered to be stable, as per the requirement of OMNR Guideline (Botzan 2011, OMNR 2002, TRCA 1994).

Soil conditions encountered at the study site can be summarized as the following:

- Topsoil about 200 mm in depth, overlaying a brown sandy silt to silty fine sand deposit extending to depth about 2 m, was observed.
- Light brown Sand and Gravel, underneath the brown sandy silt/silty fine sand, was in moist condition and extended to depth about 4 m.
- Grey Sandy silt till with some clay and trace of gravel was below the sand and gravel deposit and extended to depth about 8 m.
- Silt to Silt Till with some sand/clay and trace of gravel was observe below the sandy silt till and extended to bottom of the slope.
- Phreatic surface was assumed to be on the same level of river-water surface for slope stability analysis.

Results of computer analysis indicated that Factor of Safety FS = 0.87, for the existing riverbank slope at the section in our case study area. Factor of safety equal to 0.87 is less than the criterion required by Ontario Ministry of Natural Resource (OMNR). As a result, the existing slope cannot be considered as a long-term geotechnical stable slope (Fig. 13).

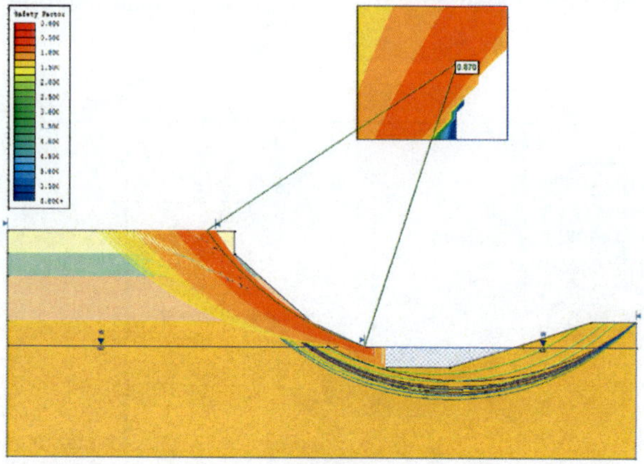

Fig. 13. Stability analysis for a section of existing slope, using Rocscience Software Slide 6.0, indicated Factor of Safety FS = 0.87 (Analyzed by Ron Xia 2019).

Back calculations were carried out on purpose of searching for long-term stable safe slope (LTSSS). With Factor of Safety equal/greater than the required criteria of FS = 1.5 for long-term stable slope, the theoretical calculated inclination would be about 2H: 1V (Fig. 14).

However, we understand that the sandy silt and sandy silt to sil strata is more erodible and a stable gradiant of 2.5H: 1V is moreh likely. Top table land of slope outside of the 2.5H: 1V inclination line would be considered as a currently long-term stable top of slope (LTSTOS).

3.2 Toe Erosion Allowance

Where the bank slope is over-steepened or subject to active toe erosion, development should be set back farther from the top of the stable slope line to ensure the development is safe from erosion and slope failure in the long term.

To determine the appropriate erosion setback for river and stream systems, engineers have to consider: 1) if the toe of the slope adjacent to the river erodes and weakens the bank, increasing the risk of slumping by active erosion, 2) the distance between the watercourse and the base of the valley wall or bank slope is 15 m or less, 3) the encountered soil types and hydraulic conditions including bank full flow velocity and the competent flow velocity of the river/stream.

A table of minimum toe erosion allowance - where river is within 15 m of slope toe (OMNR 2001) is always be applied in the erosion hazard assessment (Table 1). To

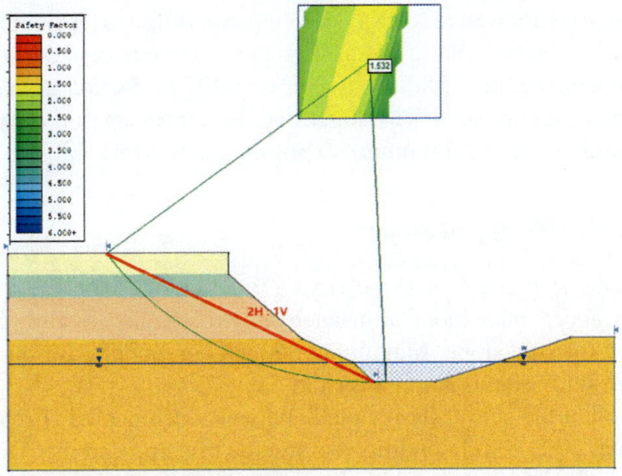

Fig. 14. Long-term stable slope back calculated, using Rocscience Software Slide 6.0, with inclination 2H: 1V (Analyzed by Ron Xia, 2015).

Table 1. Minimum toe erosion allowance – where river is within 15 m of slope toe (OMNR 2001).

Type of material Native Soil Structure	Evidence of active erosion* or where the bankfull flow velocity is greater than competent flow velocity	No evidence of active erosion		
		bankfull width		
		< 5 m	5-30 m	> 30 m
Hard rock (e.g. granite)	0 – 2 m	0 m	0 m	1 m
Soft rock (shale, limestone), cobbles, boulders	2 - 5 m	0 m	1 m	2 m
Clays, clay-silt, gravels	5 – 8 m	1 m	2 m	4 m
Sand, silt	8 – 15 m	1 – 2m	5 m	7 m

the research site, the river is within/less than 15 m from the toe. Active toe erosion is observed and slope sliding happened in the past two years. To the encountered silt to silt till with clayey/sandy matrix, trace of gravel, the expected erosion allowance would be in the range from 8–15 m, based on the information and conditions listed in the assessment Table from OMNR 2001.

3.3 Erosion Access Allowance

Erosion access allowance, or the setback needed is to ensure there's a large enough safety zone for people and vehicles to enter and exit an area during an emergency, such as a slope failure or flooding (OMNR 2001). By this way it provides a route for machinery to undertake periodic repairs as well as emergency vehicles.

The erosion access allowance is always applied in addition to the flooding hazard limit on river and stream systems and to ensure for: 1) available access during emergencies, 2) regular maintenance or repair failed structures and 3) protection from external events that affect an erosion prone area. The suggested minimum erosion allowance for river and stream systems should be minimum six (6) metres (OMNR 2001).

3.4 Long-Term Stable Top of Slope

Limit of long-term stable top of slope (LLTSTOS) can be calculated as the sum of toe erosion allowance, stable slope allowance and erosion access allowance, measured from the existing toe of a slope. Magnitude of each component, for the Thames River bank-slope at the study site, is summarized in Table 2.

Active erosion at the study site is noted and well documented (Brenk 2015; Brenk 2014; News Local 2009). Soil conditions at the toe of slope and the river bed adjacent to the toe of bank slope, is consisted of silt to silt till with sand/clay matrix and trace of gravel.

Considerable amount of soil at the slope toe is falling/sliding from upper zone of the slope in loose/soft and moist to wet conditions. Using Table 1, adopted from OMNR erosion assessment as a reference with our decades of practical experience, the expected toe-erosion allowance would be likely in the range from 8 to 15 m.

Slope is less of protection, no trees, no grasses, and no vegetation cover on the bank nearby the stream water. Plus the project site is located at the concave side of the tortuous-meanders river section. Current of lateral circular flow, from surface of water to bottom of riverbed, takes soil from the bank of concave side as sediment flow and makes deposit at convex side (Fig. 15).

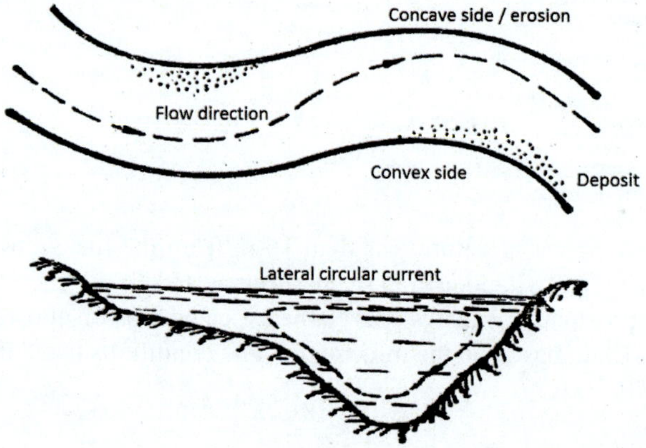

Fig. 15. Lateral circular current takes soil from concave side and deposits soil at convex side of river bank.

When the geometric sum of flow velocities, direct current and lateral current, is over/greater than the competent flow velocity, erosion will take place. Competent flow velocity v_c is the flow which the bed soil/material in the channel can sustain without erosion or scour, and it can be estimated as

$$v_c = cR^{\alpha} \tag{6}$$

Where c and α are coefficients regarding sediment deposit, soil type and erosion start up (Xu et al. 1990), R is hydraulic radius and defined as

$$R = \frac{A}{x} \tag{7}$$

Where A is the computed section area of discharge.
x is the wetting perimeter, and

$$x = b + 2h\sqrt{1 + m^2} \tag{8}$$

here b is the bottom width, h is the height of the bank submerged in water and m is the inclination of the bank slope.

The circular current is a core factor of creating erosion by under cutting. This is a common happened to all the tortuous meander rivers national and international (Xu et al. 1990). This particular is encountered at our study site and noticed during our site visiting. As a result, the most practical erosion allowance of 15 m set back should be anticipated and reasonable.

4 Lessons Learned

The OMNR Guideline provides tabulated toe erosion allowance ranges (Table 1) based on the soil type encountered at the respective location. The main criterion used in this estimation is the horizontal distance between the 'edge of water' and existing slope toe, or floodplain width.

For distances greater than 15 m, the Guideline provision is a zero for toe erosion allowance. In other words, it is considered that a slope whose toe is over 15 m away from the 'edge of water' would not be eroded in a time horizon of 100 years. It is therefore assumed that the average erosion rate of migration of the channel bed towards the respective bank is less than or equal to 0.15 m per year (Botzan 2011).

Building permit was issued in 1969 and in same year the house built up on top of the bank slope overlooking the Thames River. The couple of Shuttleworths bought the property and moved into the house at the year of 2001, which is about 32 years after, and their back yard was about 12 m deep if measured from the house to the top line of the bank slope (Brenk 2014).

Back calculation using the average erosion rate 0.15 m per year, the expected length of erosion within the past 32 years would be about 4.8 m. By this way, the original back yard could estimated as 4.8 + 12 = 16.8 m logically. Compare with the required total set back, the house is about 34 m beyond the boundary/limit of long-term stable top of slope (Table 2).

Table 2. The calculation of the long-term stable top of slope set back of the study project site.

Items	(m)
Toe erosion allowance	15
Stable slope allowance	30
Erosion access allowance	6
Limit of long-term stable top of slope	51

Frequently, the transition of toe-erosion making slope from unstable to collapse is very quietly, unnoticed until a significant proportion of soil mass from the bank slope has been removed. A large part of the rear yard collapses happened on 2009. The 40 ft deep back yard turned to 20 ft, lost about 50%. More importantly, erosion was continuing, the house is unsafe, and the couple moved out in 2012 with no choice. Again, a massive sliding happened on March 13 and 14, 2014, the house is condemned and reluctantly demolished on February 17, 2015 (Brenk 2015).

5 Conclusion

Generally, development should not occur on or on top of valley walls because the long-term stability of the slope, and therefore public health and safety, cannot be guaranteed. Development should be set back from the top of valley walls far enough to avoid increases in loading forces on the top of the slope, changes in drainage patterns that would compromise slope stability or exacerbate erosion of the slope face, and loss of stabilizing vegetation on the slope face (OMNR 2001).

Where the valley wall is over-steepened or subject to active toe erosion, development should be set farther back from the top of the valley wall so that the development will also be safe from erosion and slope failure in the long term (OMNR 2001).

The OMNR has developed a guideline to support the LTSTOS assessment. Three components introduced in the guideline are of interest in what the assessment of hazard limit is concerned: 1) toe erosion allowance, 2) stable slope allowance and 3) erosion access allowance.

Erosion access allowance provides a route for machinery to undertake periodic repairs as well as emergency vehicles. Geotechnical investigation and slope stability analysis against a Factor of Safety of minimum 1.5 are required to enable a review/issue permit of a hazard limit.

Protection structures, tress and vegetation covers always help to mitigate some hazards from toe erosion, surface downcutting/sheet erosion. Care must be taken to ensure the structures do not create new hazards and do not cause environmental damages or destroy natural systems that protect other areas.

Protection structures has to cost affordable; several pricey solutions were offered up by the municipality when it was first discovered that groundwater was causing the rapid erosion along the Thames. Unfortunately, cost of the protection construction and future maintenance could not afford by the property owners and the taxpayers and turning to the result of the house was demolished.

Hence, following the guideline of OMNR and the regulations of long-term stable top of slope (LTSTOS) assessment of local conservation authorities is a right/safe way to avoid an unexpected natural hazards/disaster of massive slope sliding.

Acknowledgements. The authors wish to express their appreciations to McClymont & Rak Engineers, and Chang'an University, for their support provided to the research team. Many thanks to Ms. Libin Xue, Mr. Ladislav Rak, and Dr. Tudor Botzan for their encourage and valuable suggestions, provided to the research team.

References

Botzan, T.: Lessons from application of environmental policies regulating development. In: Proceedings of 2011 Pan-Am CGS Geotechnical Conference, Toronto, Ontario, Canada (2011)

Brenk, D.V.: Couple's Battle with Township, County over Eroding Property Heads to Court. The London Free Press, News Local, London, Ontario (2015)

Brenk, D.V.: Home Sliding into Thames River. The London Free Press, News Local, London, Ontario (2014)

Canadian Geotechnical Society 2006: Canadian Foundation Engineering Manual, 4th edn, p. 195/488. The Canadian Geotechnical Society c/o Bitech Publishers Ltd. (2006)

Farouki, O.T.: Thermal Properties of Soils. U.S. Army CRREL Monograph 81-1, 136 p. (1981)

Hagerty, T.P., Kingston, M.S.: The Soils of Middlesex County, Volume 1. Report no. 56 of the Ontario Centre for Soil Resource Evaluation. Resources Management Branch of Ontario Ministry of Agriculture and Food, Guelph, Ontario (1992)

Hansen, M.: Photograph taken on Wednesday, April 17, 2013. The London Free Press/QMI Agency, London, Ontario (2013)

Ministry of the Environment and Ministry of Natural Resources 1975: Water Management Study for Thames River Basin, Province of Ontario, Canada (1975)

News Local 2009: Erosion Threatens Three Properties in Caradoc, Wednesday, June 10, 2009. News Local, London, Ontario (2009)

Nirupama, N., Simonovic, S.P.: Increase of flood risk due to urbanisation: a Canadian example. Nat. Hazards **40**, 25–41 (2007). https://doi.org/10.1007/s11069-006-0003-0

Ontario Ministry of Natural Resource 2001: Understanding Natural Hazards. Queen's Printer for Ontario. Printed in Ontario, Canada (2001)

Ontario Ministry of Natural Resources 2002: Technical Guide for River and Stream Systems: Erosion Hazard Limit, Province of Ontario, Canada (2002)

Patis, A.: No Solution for Owners of Condemned Homes. News.com, Blackburn (2014)

Statistics: Delaware, Ontario - The Weather Network (2015). www.theweathernetwork.com

Thames River Background Study Research Team 1998: The Thames River Watershed: A Background Study for Nomination under the Canadian Heritage Rivers System. The Upper Thames River Conservation Authority, London, Ontario, Canada (1998)

Thames River Implementation Committee 1975: Summary and Recommendations from the Thames River Basin Management Study. Thames River Implementation Committee (TRIC), London, Ontario (1975)

Toronto and Region Conservation Authority 2007: Planning and Development Procedural Manual. TRCA, Toronto, Ontario, Canada (2007)

Toronto and Region Conservation Authority 1994: Valley and Stream Corridor Management Program. TRCA, Toronto, Ontario, Canada (1994)
Upper Thames River Conservation Authority 1998: The Thames River Watershed, A Background Study for Nomination under the Canadian Heritage Rivers System. UTRCA, London, Ontario, Canada (1998)
Woo, M.K., Xia, Z.J.: Suprapermafrost groundwater seepage in gravelly terrain, Resolute, NWT, Canada. J. Permafrost Periglac. Process. **6**, 57–72 (1995a)
Woo, M.K., Xia, Z.J.: Effects of hydrology on the thermal conditions of the active layer. J. Nord. Hydrol. **27**, 129–142 (1995b)
Xia, Z.J., Xue, L., Shahine, J., Sun, Y.: Theoretical analysis of slope erosion. In: Proceedings of 55th Canadian Geotechnical Conference and 3rd Joint IAH and CGS Groundwater Conference, Niagara Falls, Ontario, Canada (2002)
Xia, Z.J., Woo, M.K., Chahine, J., Sun, Y.F.: Computation of frost table profile using Green Theorem. In: Proceedings of 54th Canadian Geotechnical Conference, 2nd Joint IAH and CGS Groundwater Conference, Calgary, Canada (2001)
Xia, Z.J.: A study of thermal cracks in frozen ground no. 3. In: Fourth International Symposium on Ground Freezing, Sapporo, 5–7 August 1985 (1985)
Xu, Y.Ch., Hu, D.B., Xue, C.Y.: Hydraulics, 3rd edn. Textbook of Dalian University of Science and Technology, Science Press, Beijing (1990)

Characteristics of Recycled Micro Powder Produced Using Construction Waste

Jinjin Shi[1,2,3(✉)], Miao Xu[1,3], Yingbiao Wu[1,2,3], and Jinyan Liu[1,2]

[1] Cangzhou Municipal Engineering Company Limited, Cangzhou 061000, China
[2] Hebei Province Road Materials and Technology Engineering Technology Research Center, Cangzhou 061000, China
[3] Hebei Industrial Technology Research Institute of Construction Waste Comprehensive Utilization, Cangzhou 061000, China

Abstract. Six different recycled micro powders were prepared by grinding the construction wastes that composed of concrete, tile, brick, and concrete. Experimental researches on their physical and chemical properties, particle size distribution, chemical composition and strength activity index were investigated and compared with the characteristics of cement, slag powder and fly ash. The results suggest that the specific surface area of the recycled micro powder prepared by grinding three types of construction waste for more than 30 min would reach more than 500 m^2/kg. When the specific surface area of the recycled micro powder is $\geq$500 m^2/kg, its density, water demand ratio, and other conventional physical and chemical properties and activities are not affected by fineness. The content of CaO is higher and SiO_2 is lower in the chemical components of the recycled micro powder of waste concrete, while the content of CaO is lower and SiO_2 is higher in the chemical components of recycled micro powder of waste brick and tile. The 28-day activity of the recycled micro powder meets the specification requirements of fly ash.

1 Preface

Recycled micro powder is a kind of fine powder which can be used as mineral admixture of cement concrete, waste brick, and tile after grinding to a certain degree of fineness. It is generally defined that particle size should be less than 0.16 mm [1] in order to induce its filling and activity effects [2]. The recycled aggregate of construction waste is a new way of recycling construction waste, which can significantly improve the economic added values and engineering properties of construction waste. For this study, six different recycled micro powders of construction waste were prepared, and their basic properties were studied. Compared with those of cement, slag powder, and fly ash, the feasibility of using them in cement concrete are presented in this paper.

2 Selection of Raw Materials and Preparation of Regenerated Powder

P.O.42.5 cement of Tang xian Ji dong Cement Co., Ltd., S95 slag powder of Tangshan "Xing wang" brand and class-F class-II fly ash produced by De zhou Power Plant were

A. Tapase et al. (Eds.): GeoChina 2021, SUCI, pp. 167–176, 2021.
https://doi.org/10.1007/978-3-030-79644-0_13

used as test materials. The technical performance indices of these materials met the requirements of relevant specifications [3–5].

The construction wastes included concrete, bricks, and tiles were derived from demolitions of concrete brick and buildings. The construction wastes were broken into recycled aggregates with particle size of 5–10 mm and washed several times to remove substances and debris. The basic performance indices of the recycled aggregates were tested according to the requirements of *Recycled Coarse Aggregate For Concrete* (*GB/T 25177–2010*) [6]. The test results are shown in Table 1.

Table 1. Test results of recycled aggregate raw materials of 5–10 mm waste concrete, and waste brick, and tile mixed construction waste

Test items	Test result	Technical index [6]		
		Class I	Class II	Class III
apparent density (kg/m^3)	2410	>2450	>2350	>2250
Void fraction (%)	48	<47	<50	<53
Crushing index (%)	19.6	<12	<20	<30
Mud content (%)	0.86	<1	<2	<3
Content of needle and flake particles (%)	2.3		<10	
Impurity content (%)	0.2		<1	

Note: mass ratio of recycled aggregates of waste concrete, waste brick, and tile is about 7:3

The recycled aggregates mainly compose of concrete, brick, and tile. SM-500 cement test mill (mill speed: 48 rpm) was used to grind these three kinds of recycled aggregates for 30 min or 1 h. Therefore, the test metric included six kinds of recycled micro powders, which were respectively labeled as 1#–6# recycled micro powders. The main components and grinding times are shown in Table 2.

Table 2. Main components and grinding time of recycled micro powders

Sample name	Composition	Grinding time
1# Recycled micro powder	Waste concrete	30 min
2# Recycled micro powder	Waste concrete	1 h
3# Recycled micro powder	Waste bricks and tiles	30 min
4# Recycled micro powder	Waste bricks and tiles	1 h
5# Recycled micro powder	Mixing waste concrete with waste brick and tile	30 min
6# Recycled micro powder	Mixing waste concrete with waste brick and tile	1 h

3 Test Method

General Portland Cement (GB 175-2007) [3], *Granulated Blast Furnace Slag Powder For Cement And Concrete* (*GB/T 18046-2008*) [4] and *Fly Ash For Cement And Concrete* (*GB/T 1596-2017*) [5] standards and specifications were used to test the conventional physical and chemical properties of the recycled micro powders and other mineral admixtures.

A laser particle size analyzer was used to compare and analyze the particle size distribution of the construction waste recycled micro powders and other mineral admixtures in accordance to the test method of *Particle Size Analysis Laser Diffraction Method* (*GB/T 19077-2016*) [7]. The particle size distribution was measured following the physical phenomenon that particles can cause laser scattering. As shown in Fig. 1, the scattering light I_1 is caused by larger particles, while the scattering light I_2 is caused by smaller particles. The average particle size of a powder is usually expressed by D_{50} and D_{90}. D_{50} is the corresponding particle size when the cumulative percentage on the particle size distribution of a sample reaches 50%. D_{90} is the corresponding particle size when the cumulative percentage on the particle size distribution of a sample reaches 90%.

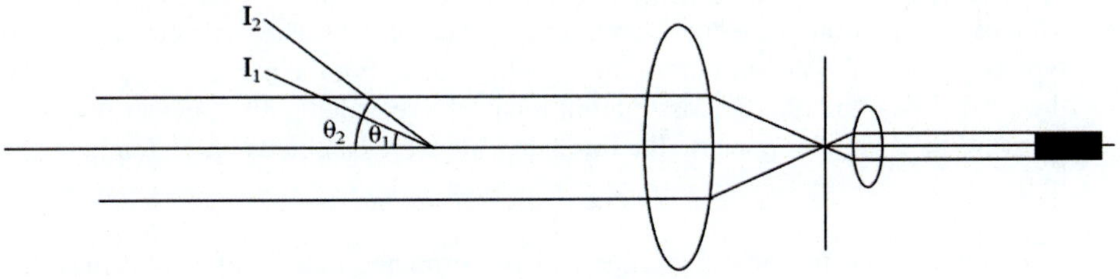

Fig. 1. Schematic diagram of laser particle sizer test

According to the *Classification, Safety Requirements and Test of Portable Tube Excited X-ray Fluorescence Analyzer* (*GB/T 35734-2017*) [8], the chemical composition of a recycled micro powder of the construction waste and other mineral admixtures were analyzed by X-ray fluorescence spectrometer. The X-ray fluorescence spectrometer consists of excitation source (X-ray tube) and detection system. The X-ray tube generates incident X-ray (primary X-ray). Each element in the excited sample emits secondary X-rays, and the secondary x-rays emitted by different elements have specific energy characteristics or wavelength characteristics. The instrument measures the energy and quantity of these emitted secondary X-rays. Then, the software converts the information collected by the instrument into the types and contents of various elements present in the sample.

According to the test method of fly ash strength activity index in Appendix C of *Fly Ash Used In Cement And Concrete* (*GB/T 1596-2017*) [5], the strength activity index tests of the recycled micro powders of the construction waste and other mineral admixtures were carried out, and the results are presented in the subsequent sections.

4 Result Analyses

4.1 Conventional Physical and Chemical Properties

The physical and chemical properties of 1#–6# recycled micro powders, cement, fly ash, and slag powder were studied. The test results are presented in Table 3. These results indicate that:

(1) The specific surface area of recycled micro powder prepared by grinding recycled aggregate for 30 min is more than 500 m^2/kg, which is greater than that of cement, slag powder, and fly ash.
(2) With increasing grinding time, specific surface area, and density, water content and water demand ratio of the recycled micro powder are increased slightly.
(3) The density of recycled micro powder is about 2.6 g/cm^3, which is between fly ash and slag powder.
(4) The water requirement ratio of the recycled micro powders prepared by different components of the recycled aggregates has little difference. However, the ratio is larger than that of fly ash or slag powder. This is due to the large specific surface area and rough particle surface of the recycled micro powder. As a result, the water demand of mixture blended with the recycled micro powder increases [9].
(5) The loss on ignition of the recycled micro powder of the waste concrete and mixed type is greater than that of other mineral admixtures. For the recycled aggregate of the waste brick and tile, the loss on ignition of the recycled micro powder prepared by grinding is smaller because the brick and tile were calcined at high temperature in the forming process.

In conclusion, the physical and chemical performance indices of 1#–6# recycled micro powders basically meet the technical index requirements of grade II fly ash in Fly Ash For Cement And Concrete (GB/T 1596-2017) [5]. Thus, they meet the requirements to use as a mineral admixture of cement concrete. Therefore, it is feasible to replace grade II fly ash with the recycled micro powders.

Table 3. Test results of physical and chemical properties

Sample name	Fineness (45 μm square hole sieve residue) (%)	Density (g/cm^3)	Comparison table area (m^2/kg)	Moisture content (%)	Loss on ignition (%)	Stability (mm)	Water demand Volume ratio
1# Recycled micro powder	19	2.61	648	0.8	15	1	104
2# Recycled micro powder	/	2.62	883	1	16	1	105

(*continued*)

Table 3. (*continued*)

Sample name	Fineness (45 μm square hole sieve residue) (%)	Density (g/cm^3)	Comparison table area (m^2/kg)	Moisture content (%)	Loss on ignition (%)	Stability (mm)	Water demand Volume ratio
3# Recycled micro powder	20	2.59	545	0.4	1.6	1	104
4# Recycled micro powder	12	2.65	843	0.4	3.5	1	106
5# Recycled micro powder	19	2.6	542	0.2	10	3	106
6# Recycled micro powder	/	2.61	961	0.4	10	3	106
Cement	/	2.99	418	0.1	1	2	/
Fly ash	22	2.35	317	0.3	1	2	98
Slag powder	/	2.84	473	0.2	0.8	/	99
Technical index of grade II fly ash [5]	≤30	≤2.6	/	≤1.0	≤8.0	≤5.0	≤105

4.2 Particle Size Distribution

The particle size analyses of 1#-6# recycled micro powders, cement, fly ash and slag powder were carried out using a laser particle size analyzer. The results are shown in Table 4 and Fig. 2, 3, 4, 5, 6, 7, 8, 9 and Fig. 10.

Table 4. Calculation results of particle size parameters

Sample name	D_{50} (μm)	D_{90} (μm)
1# Recycled micro powder	17	78
2# Recycled micro powder	11	104
3# Recycled micro powder	7.4	55
4# Recycled micro powder	6.9	92
5# Recycled micro powder	15	64
6# Recycled micro powder	9.5	105

(*continued*)

Table 4. (*continued*)

Sample name	D_{50} (μm)	D_{90} (μm)
Cement	12	35
Fly ash	19	82
Slag powder	14	35

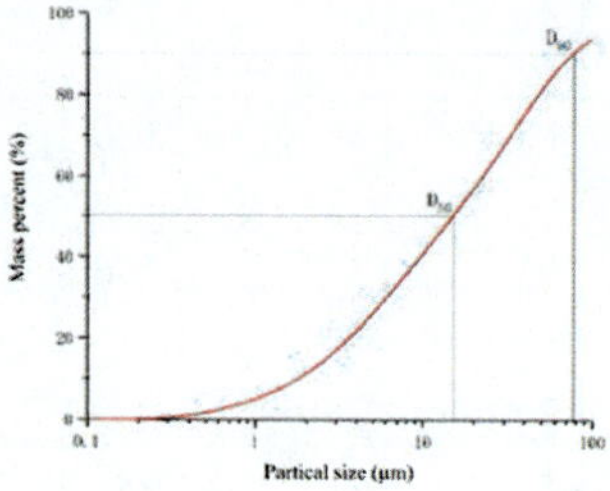

Fig. 2. 1# Recycled micro powder

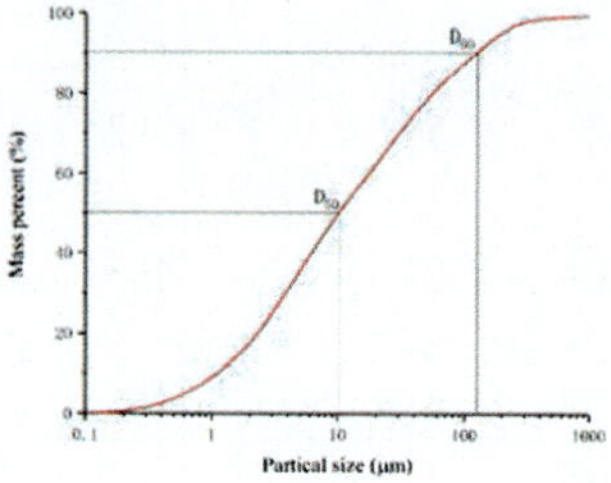

Fig. 3. 2# Recycled micro powder

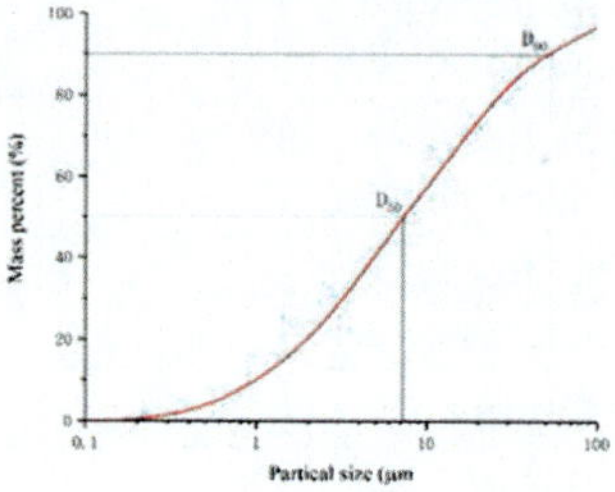

Fig. 4. 3# Recycled micro powder

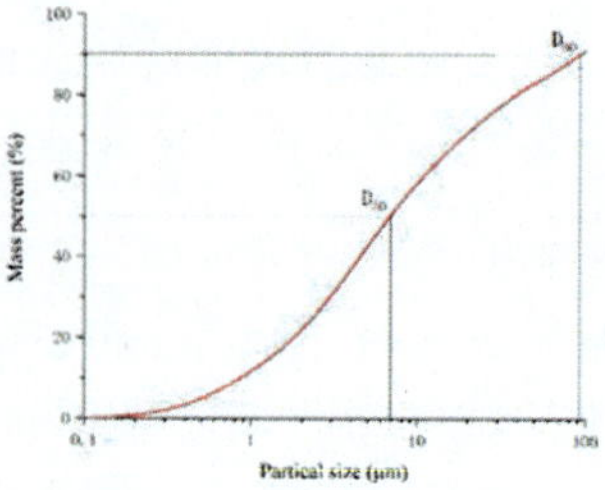

Fig. 5. 4# Recycled micro powder

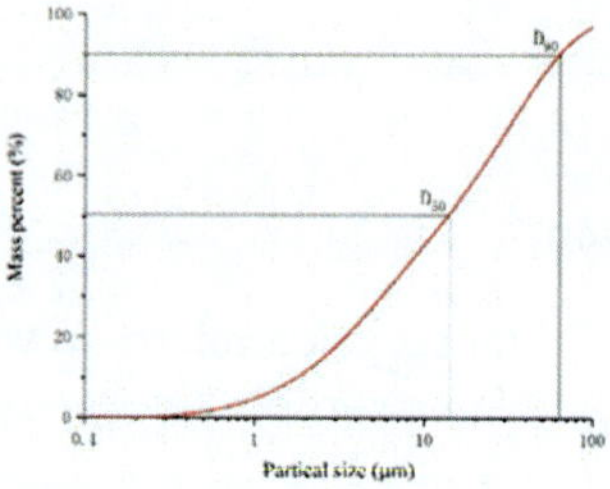

Fig. 6. 5# Recycled micro powder

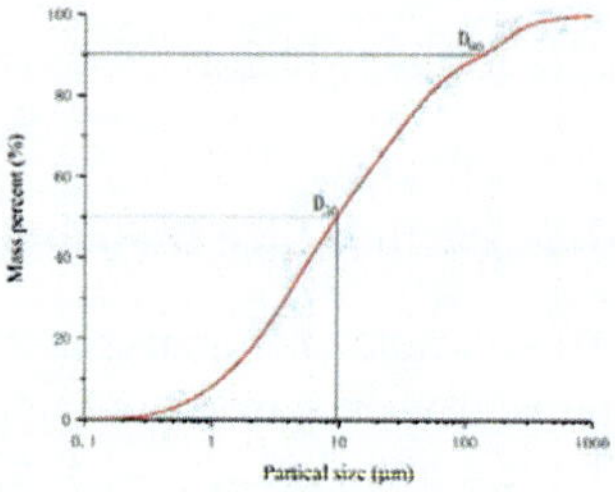

Fig. 7. 6# Recycled micro powder

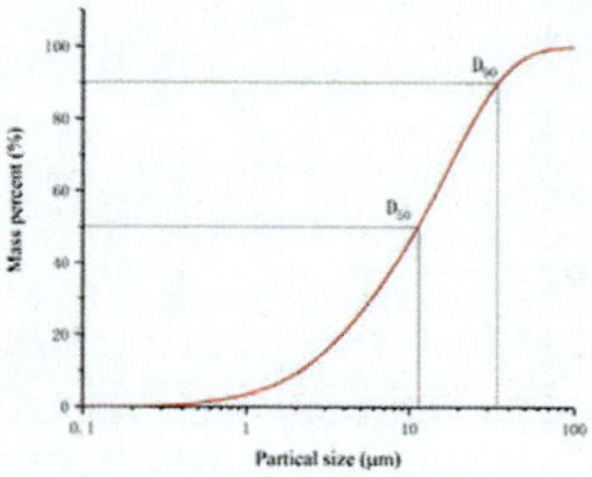

Fig. 8. Cement

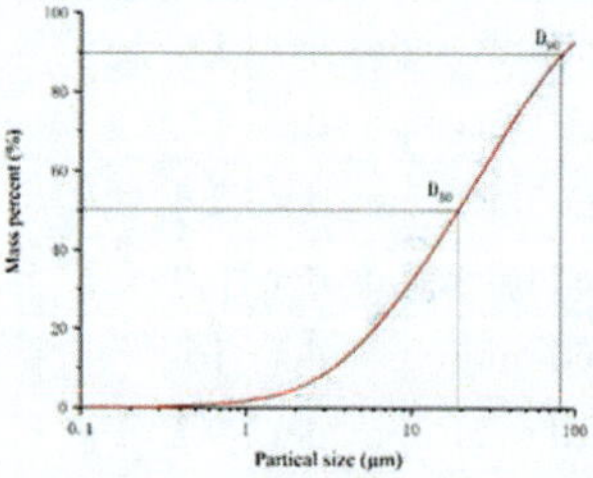

Fig. 9. Fly ash

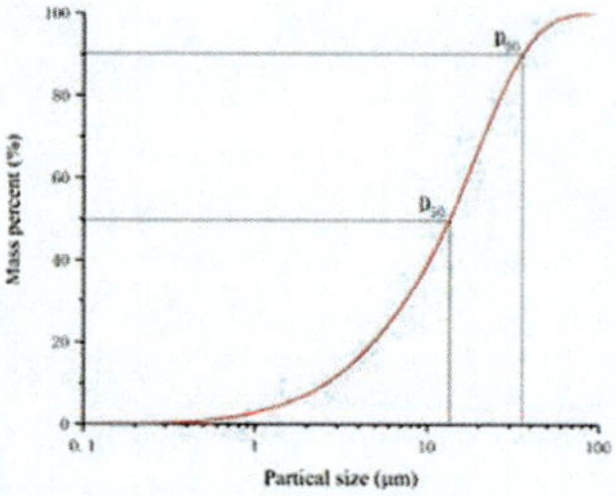

Fig. 10. Slag powder

The experimental results show that:

(1) The average particle size (D_{50}) of the waste concrete and the mixed recycled fine powder is not significantly different from cement, slag powder, and fly ash, while the average particle size (D_{50}) of the waste brick and tile recycled fine powder is significantly smaller than other mineral admixtures. The recycled waste brick and tile aggregates are easier to grind due to their lower hardness. Therefore, the average particle size should be smaller under the same grinding time.
(2) As the grinding time prolonged, the D_{50} of the same type of recycled micro powder decreases, but D_{90} increases. When the recycled fine powders of construction wastes are ground to a certain level (specific surface area $\geq$843m^2/kg), more micro-cracks are formed on the surface of the particles. When the diameter distribution is uneven, due to characteristics of polarization between coarse and fine particles, it leads to a decrease in the D_{50} of the regenerated fine powder, but an increase in the D_{90}.

4.3 Chemical Composition

The chemical compositions of 1#–6# recycled micro powders, cement, fly ash, and slag powder were measured by X-ray fluorescence spectrometer. The test results are shown in Table 5.

Table 5. Chemical composition of each admixture

Sample name	Chemical composition content (%)							
	Al_2O_3	SiO_2	CaO	Fe_2O_3	MgO	Na_2O	K_2O	SO_3
1# Recycled micro powder	12.76	46.81	24.26	4.47	5.75	1.86	2.23	/
2# Recycled micro powder	12.45	46.53	23.43	5.50	6.21	1.84	2.25	/
3# Recycled micro powder	16.43	58.60	8.17	7.85	2.77	2.09	2.66	/
4# Recycled micro powder	15.68	57.51	9.27	8.18	3.09	2.16	2.60	/
5# Recycled micro powder	13.93	48.83	21.58	4.43	5.16	1.85	2.35	/
6# Recycled micro powder	13.79	49.99	20.22	4.49	5.26	2.02	2.41	/
Cement	6.81	20.46	59.11	2.94	4.61	/	1.40	3.22
Fly ash	40.82	46.94	3.35	4.40	/	/	/	/
Slag powder	19.20	31.04	32.93	1.25	8.61	/	/	3.63

It can be seen from the data in Table 5 that the main chemical compositions (Al_2O_3, SiO_2, Cao, Fe_2O_3, and MgO) are basically the same, but the relative contents are different.

The relative contents of main chemical components in the cement and slag powder are similar, in which CaO content is the highest, followed by SiO_2. The content of SiO_2 in fly ash is high, but its CaO content is very low.

Comparing the chemical composition of the recycled micro powders of the waste concrete and the waste brick and tile, it can be observed that the content of Cao and SiO_2

in the recycled micro powder of the waste concrete is high, but the content of SiO_2 in the recycled powder of the waste brick and tile is low. This is mainly due to the large use of cement in concrete, and the main raw material for cement production is limestone, and the raw material for firing brick and tile is clay. Therefore, it causes the content of SiO_2 to be high. It can be seen that the type of raw materials has a great influence on the chemical composition of micro powder, which is consistent with the research conclusion in the literature [10].

The main chemical composition content of the mixed recycled micro powder is between the waste concrete and the waste brick and tile. Because of the large proportion of the waste concrete, the main chemical composition content of the mixed recycled micro powder is closer to that of fresh concrete.

The relative content of the main chemical components of the recycled micro powders of the waste concrete is close to that of the cement and slag powder. Their activity characteristics may be similar to that of the cement and slag powder.

4.4 Strength Activity Index

The strength activity indices of 1#–6# recycled micro powders, cement, fly ash and slag powder were tested. The test results are shown in Table 6.

Table 6. Test results of strength activity index

Sample name	7-day strength activity index (%)	28-day intensity activity index (%)
1# Recycled micro powder	46	70
2# Recycled micro powder	50	69
3# Recycled micro powder	51	81
4# Recycled micro powder	55	78
5# Recycled micro powder	47	71
6# Recycled micro powder	51	71
cement	100	100
Fly ash	/	74
Slag powder	88	95
Technical index of fly ash [5]	/	≥70

The experimental results show that:

(1) The 7-day and 28-day activity indices of the six recycled micro powders can basically meet the technical requirements of fly ash. But the activity indices fall short of S95 grade slag powder.
(2) Under the same grinding time, the 7-day and 28-day activity indices of the recycled micro powder of the waste brick and tile are slightly higher than that of waste concrete.

(3) Prolonging the grinding time can improve the 7-day activity indexes of the recycled powder. However, it has no effect on the 28-day activity.

To sum up, the activity indices of the three construction wastes (recycled micro powders) can meet the technical requirements of fly ash. The grinding time has a certain limitation on the improvement of the recycled micro powder. For practical application, the grinding time of the recycled construction waste aggregate should not exceed 30 min.

5 Conclusions

(1) The specific surface areas of recycled micro powder can reach more than 500 m^2/kg by grinding 5–10 mm recycled aggregate of construction wastes for 30 min. The specific surface areas are larger than that of cement, slag powder and fly ash.
(2) With the increase of specific surface areas, the conventional physical and chemical properties changed little, such as density and water demand ratio of the same type of recycled micro powder. For practical application, the grinding time of the recycled construction waste aggregate should not exceed 30 min.
(3) Under the same grinding time, the activity index of recycled micro powder of waste brick and tile is slightly higher than that of waste concrete. When the recycled micro powder was grinded to a certain extent (specific surface area $\geq$843 m^2/kg), the particle agglomeration phenomenon would appear, resulting in uneven distribution of particles.
(4) The relative content of the main chemical components of the recycled powder of waste concrete is similar to that of cement and slag powder.
(5) The results show that the strength indexes of recycled micro power of construction waste basically meet the standard of grade II fly ash. Thus, the recycled micro power is feasible to replace fly ash as mineral admixture of cement concrete

References

1. Wang, S., Qian, N.: The application research status of construction waste regenerated powder in mortar. China Hous. Facil. (03), 82–83+118 (2019)
2. Li, J.: Explore the basic properties and application of construction waste recycled micropowder. Build. Mater. Decoration (18), 46–47 (2020)
3. GB 175-2007: Common Portland Cement. China Standard Press (2007)
4. GB/T 18046-2008: Ground Granulated Blast Furnace Slag Used for Cement and Concrete. China Standard Press (2008)
5. GB/T 1596-2017: Fly Ash Used for Cement and Concrete. China Standard Press (2017)
6. GB/T 25177-2010: Recycled Coarse Aggregate for Concrete. China Standard Press (2010)
7. GB/T 19077-2016: Particle Size Analysis—Laser Diffraction Methods. China Standard Press (2016)
8. GB/T 35734-2017: Portable Tube-Excited X-Ray Fluorescence Analysis Equipment—Classification, Safety Requirements and Test. China Standard Press (2018)

9. Zhao, L.: Study on basic properties and application of recycled micropowder from construction waste. Beijing Jianzhu University (2019)
10. Lan, C., Luo, J., Lu, J., Chen, J., Yang, S.: Comparative test study on the performance of recycled micropowder. Concr. Cem. Prod. (10), 110–112 (2018)

Simulation of Rock Hydraulics in Rock Joint by Using Discrete Element Method

C. C. Chiu[1] (✉) and M. C. Weng[2]

[1] Institute of Mineral Resources Engineering, National Taipei University of Technology, Taipei, Taiwan
ccchiu@ntut.edu.tw

[2] Department of Civil Engineering, National Yang Ming Chiao Tung University, Hsinchu, Taiwan

Abstract. This study proposed a method to calculate rock hydraulics in rock joint by merging the Lattice method and the particulate interface model in the particulate discrete element method. The Lattice method builds a Voronoi-cell structure based on particles' contact, which can be used to calculate fluid transmission between cells; the particulate interface model can modify the results of rock joint simulation in the discrete element method. By combining the above two methods, correct mechanical behavior can be obtained by the particulate interface model, and the fluid transmission in the rock joint can be calculated by using the Lattice model. The theory of the Lattice model and the particulate interface model is first introduced. To validate the correctness of the proposed model, the simulated water head and the variation of flow rate in a single fracture under un-shear and shear conditions are validated by the analytical solution. Finally, a re-meshing technique to handle the variation of pore mesh has been proposed to support the operation of the modified fluid flow algorithm during shearing.

Keywords: Discrete element method · Fluid transmission · Fracture · Rock hydraulics · Seepage

1 Introduction

It is an important issue to evaluating fluid transmission in a jointed rock mass in rock mechanics. Lots of civil engineering research, such as tunneling inflow evaluation (Shin and Santamarina 2019), geothermic exploration (Yoon et al. 2013) and assessment of the performance of nuclear waste disposal required the aid of rock hydraulic theory. The fracture-oriented models is a modern technique to analyse the flow behaviour in rock mass. In fracture-oriented models, the distribution of rock joints in a rock mass is established using a discrete fracture network (DFN), and the corresponding permeability, which is calculated from the DFN and the rock joint properties, is obtained to determine the permeability in the rock mass (Wang et al. 2015). However, when the rock mass suffers external stress, the induced deformation changes the aperture size and roughness of the fracture, and the original DFN model therefore cannot deal with such effects of the external stress on the fracture. To solve this problem, the discrete element method

A. Tapase et al. (Eds.): GeoChina 2021, SUCI, pp. 177–183, 2021.
https://doi.org/10.1007/978-3-030-79644-0_14

(DEM) provides a solution that makes the calculation of DFN under external stresses possible.

The particulate DEM is a powerful tool to investigate several problems in civil engineering such as tunneling, landslide analysis, and mining engineering (Lo et al. 2016). The DEM has several unique characteristics and advantages, such as good at simulates crack propagation, large deformation, block separation and block movement (Chiu and Weng 2019a). However, the application of the particulate DEM to fluid transmission is still under development. Previous researchers have investigated fluid transmission in rock material by using a fluid cell algorithm couple with the particulate DEM (Hazzard et al. 2002; Yoon et al. 2017, 2014), but the simulation still need back calculations and not focus on simulates the fluid transmission in rock joint.

To simulate the transmission of fluid in rock joint by using the particulate DEM, this study adopts the particulate interface model (PIM) (Chiu and Weng 2019a, b) to simulate the mechanical behaviour of fractures. The PIM is an interface model build for particulate DEM, which can simulates the mechanical behaviour of a rock joint in terms of its macroscopic parameters without the need for a time-consuming back-analysis. In this study, the fluid flow algorithm of Yoon et al. (2014) is modified to incorporate the PIM and used to evaluate fluid transmission in rock joints, which we called as the lattice model. In the manuscript, the proposed model is firstly validated using the analytical solution from the field of fluid mechanics and its application is then demonstrated using several examples. The PIM, incorporating the lattice model, can be used in the particulate DEM to determine fluid transmission in fractures, providing an effective reasonable way to investigate problems in rock hydraulics.

2 Particulate Interface Model and Lattice Model

To simulate fluid transmission in rock joints, the lattice model is used with the particulate interface model (PIM), which are executed in the discrete element software Particle Flow Code 2 Dimension (PFC2D). This section introduces the structure of the PIM. Then, the analytical solution for flow through a single fracture is presented as the theoretical basis for the subsequent validation of the method. Finally, the lattice model, and its conceptual basis, are described.

2.1 Particulate Interface Model (PIM)

The PIM is used with the particulate discrete element method to simulate rock joint behaviour. It is based on the basic theory of the smooth joint model. In the PIM, a fracture consists of a series of fracture contacts in the designated plane. Those contacts are identified using the joint sides checking approach (Mehranpour and Kulatilake 2017). In the algorithm of the PIM, the overall behaviour of the fracture contacts is consistent with the behaviour of the fracture.

The algorithm of the PIM involves six modifications, which are (1) segment area equalization, (2) shear stiffness adjustment, (3) surplus force recapture, (4) shear force compensation, (5) normal force redistribution, and (6) normal stiffness adjustment. More detailed information and algorithms associated with the PIM can be found in two related studies (Chiu and Weng 2019a, b).

2.2 Lattice Model

The basic theory of the lattice model in this study is modified from that of Yoon et al. (2017). The concept was originally proposed by Cundall in an unpublished technical note; Hazzard et al. (2002) and Yoon et al. (2017) subsequently used it to construct a fluid transmission network to evaluate injection behaviour in rock. Figure 1 is the schematic diagram of the lattice model. Virtual pore space (black lines in Fig. 1) is defined that is surrounded by adjacent particles in contact. If the edge of a pore space is regarded as a fracture contact, then the pore space will be considered as part of fluid transmission route, which is the blue cells and red lines in the figure. A series of pore spaces connect to form flow channels or flow route. The fluid can only be transmitted through the channels among pore spaces. The fracture is sandwiched between two blocks. Every pore spaces can store lots of information such as pore pressure and neighbour contacts.

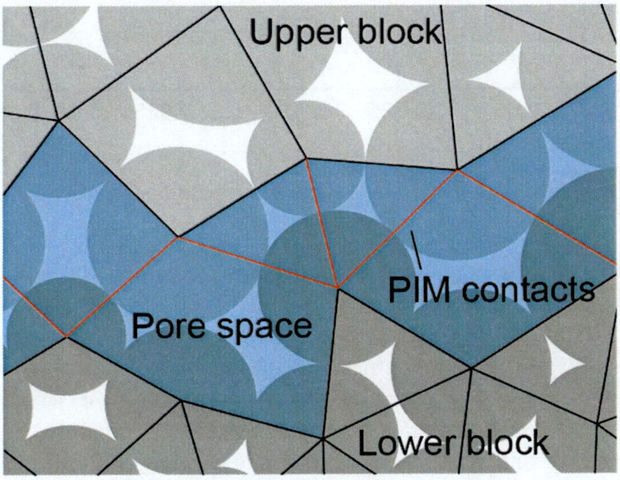

Fig. 1. Lattice model.

Figure 2 illustrates a benchmark analysis case. The figure represents fluid transmission through a fracture under different normal pressures and shear stress. The black lines form closed pore spaces in the rock matrix, indicating that no fluid transmission occurs in the intact rock. The red lines represent fracture contacts, which exhibit the mechanical behaviour of the fracture and transmit the fluid. The blue line indicates the water head distribution along the fracture during fluid transmission.

3 Validation of Fluid Transmission in a Single Fracture

3.1 Model Setting

A numerical test of fluid transmission through a fracture is performed to validate the performance of the PIM with the lattice model. Figure 2 presents a 2D case that involves a single fracture that is sandwiched between two blocks. The length of the joint and

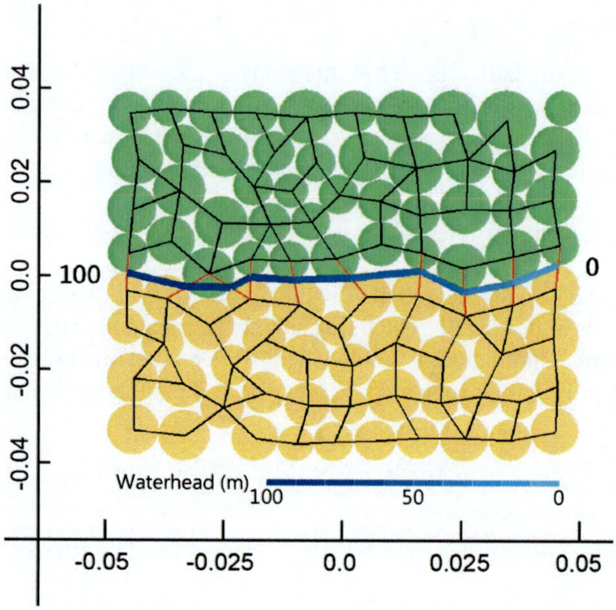

Fig. 2. Simulation of flow along a single fracture using the proposed model.

the block is 0.1 m and the height is 0.08 m. Total particle number is 86, and the mean particle radius is 0.005 m. The mechanical model of the rock joint uses Barton's shear strength criterion (Barton 1973). The required parameters are as follows; joint roughness coefficient is 14.4; joint wall strength is 7.86 MPa; basic friction angle is 31°; joint initial normal stiffness is 0.3 GPa/m; and ultimate normal closure is 1.8×10^{-4} m. The simulation contains two loading stages: for the first stage, normal loading is applied on upper box; for the second stage, the upper box is sheared and the shear stress of the specimen is counted. The simulation of fluid transmission through the fracture is firstly validated under only a normal stress of 0.5 MPa. The water head at the source (H_0, at the left end of the block in Fig. 2) is a constant boundary and is set to 100 m and the water head at the end (H_L, at the right end of the block in Fig. 2) is also a constant boundary and is set to 0 m, so the fluid is constantly being replenished and flowing out.

3.2 Water Head Distribution

Figure 3a displays the theoretical and simulated distributions of the water head at various flow times from 0.02 s to 0.5 s. For the conducting aperture under normal stress of 0.5 MPa, the hydraulic conductivity is 1.73×10^{-2} (m/s). In Fig. 3, the black lines represent the analytical distribution of the water head and the red dots show the simulated water head at the pore space nodes. The results from 0.02 s to 0.1 s show the flow transmission in the transient state, and those from 0.5 s demonstrate the flow transmission is in the steady state because it is a straight line. The simulated and theoretical water head distributions agree closely. The water head distributions under different hydraulic conductivities are also compared. The range of hydraulic conductivity are from 1.73×10^{-4} to 1.73×10^{-1} m/s, and the flow time is equally 0.1 s. Figure 3 shows the theoretical and simulated water head distributions, which two results are fit. Based on the results, the proposed model is able to simulate fluid transmission in a single fracture in both steady and transient states.

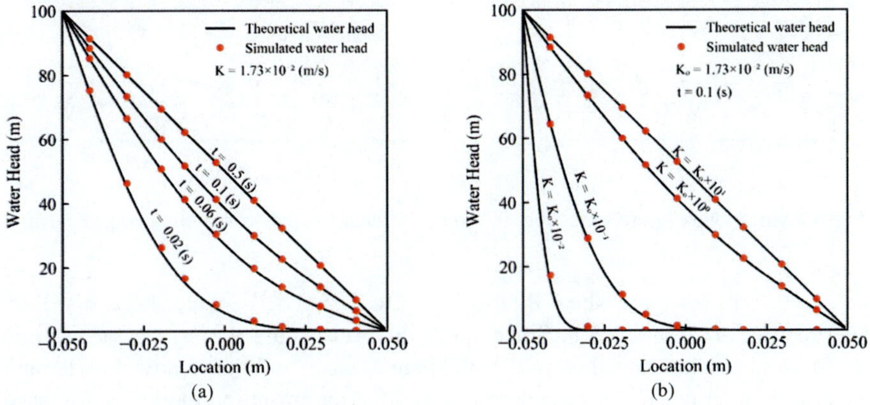

Fig. 3. Distribution of theoretical and simulated water heads for different (a) flow times (b) hydraulic conductivities.

4 Variation of Fluid Transmission During Fracture Shearing

Fluid transmission under shearing involves variation in the pore space distribution under applied shear stress, so it is required to develop an algorithm for re-meshing the pore space once for use once if the original pore space changes under shearing.

4.1 Algorithm for Re-meshing Pore Space

The re-meshing technique reconstructs and refreshes the pore space when the contact between particles is altered during fracture shearing. During shearing, if a new fracture contact forms and is located at the boundary of the fracture, then one new pore space is generated; if it is located in a existing pore space, then that pore space is divided into two new pore spaces. When an existing fracture contact vanishes, the pore space is eliminated if it is located at the boundary; otherwise, two pore spaces adjacent to the eliminated contact are combined as a single new pore space. Figure 4 plots the variation of pore space distribution in a fracture during the shearing process. The algorithm for re-meshing the pore space provides a reasonable pore space distribution when new fracture contacts are generated or existing fracture contacts vanish.

4.2 Variations of Water Head

Fluid transmission in a fracture under shearing is simulated to verify the effectiveness of the proposed re-meshing algorithm. Different to the settings of Sect. 3.1, now we apply the shear velocity of 0.01 m/s. The conducting aperture is increased from 4.6×10^{-7} m

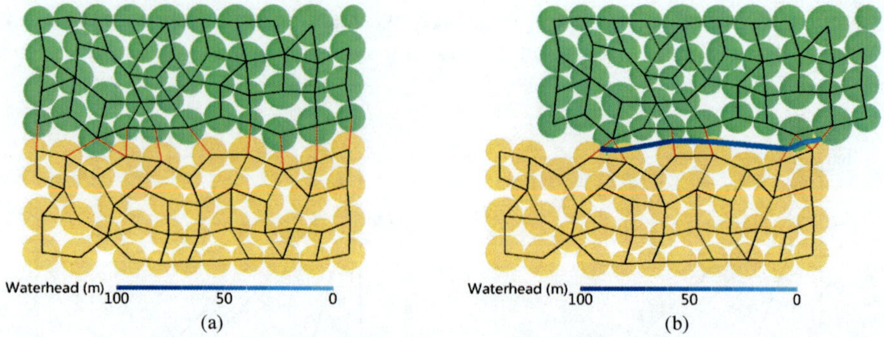

Fig. 4. Variations of pore space distribution and water head distribution during shearing.

to 4.1×10^{-5} m owing to shear dilation. Figure 5 shows the water head distribution under shearing at1.0 and 1.5 s. Since shear dilation occurs during fracture shearing, the aperture is much larger than in the un-sheared case, and the steady state is rapidly reached in both cases. The results demonstrate that the proposed model can be used to simulate fluid transmission in a fracture under shearing.

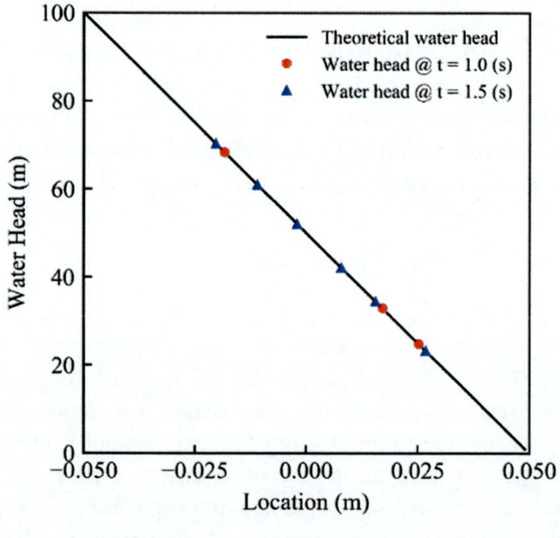

Fig. 5. Water head distribution under shearing at 1 and 1.5 s.

5 Conclusion

This study proposes the performance of particulate interface model (PIM) couple with the lattice model to simulate fluid transmission in rock fractures. In the proposed model, a series of pore spaces connect together and become a flow channel, which exists between two rock blocks, so the fluid can only be transmitted in the fracture rather than in the rock pores. A series of numerical tests of fluid transmits through a fracture under different normal stresses are conducted to validate the performance of the PIM with the lattice model. Based on the results, the proposed model can be used to simulate fluid transmission in a single fracture in both steady and transient states. Additionally, to consider fluid transmission under shearing in the DEM, we proposes a re-meshing algorithm to deal with variations in the pore space during joint shearing. The above results show that the proposed model can provide a reasonable simulation of the water head distribution and flow rate in a general rock joint.

Acknowledgements. The authors would like to thank the Ministry of Science and Technology, Taiwan, for financially supporting this research under contract MOST 108-2218-E-027-016-MY2.

References

Barton, N.: Review of a new shear-strength criterion for rock joints. Eng. Geol. **7**, 287–332 (1973). https://doi.org/10.1016/0013-7952(73)90013-6

Chiu, C.C., Weng, M.C.: DEM simulation of planar sliding using a particulate interface model considering velocity-dependent friction. Comput. Geotech. **112**, 51–59 (2019a). https://doi.org/10.1016/j.compgeo.2019.04.001

Chiu, C.C., Weng, M.C.: Simulating interface characteristics by using a particulate interface model of a discrete element method. Comput. Geotech. **109**, 1–11 (2019b). https://doi.org/10.1016/j.compgeo.2019.01.011

Hazzard, J.F., Young, R.P., Oates, S.J.: Numerical modeling of seismicity induced by fluid injection in a fractured reservoir. In: 5th North American Rock Mechanics Symposium, Toronto, Canada, pp. 1023–1030 (2002)

Lo, C.-M., Li, H.-H., Ke, C.-C.: Kinematic model of a translational slide in the Cidu section of the Formosan Freeway. Landslides **13**(1), 141–151 (2016). https://doi.org/10.1007/s10346-015-0650-x

Mehranpour, M.H., Kulatilake, P.H.S.W.: Improvements for the smooth joint contact model of the particle flow code and its applications. Comput. Geotech. **87**, 163–177 (2017). http://dx.doi.org/10.1016/j.compgeo.2017.02.012

Shin, H., Santamarina, J.C.: An implicit joint-continuum model for the hydro-mechanical analysis of fractured rock masses. Int. J. Rock Mech. Min. Sci. **119**, 140–148 (2019). https://doi.org/10.1016/j.ijrmms.2019.04.006

Wang, T.T., Zhan, S.S., Huang, T.H.: Determining transmissivity of fracture sets with statistical significance using single-borehole hydraulic tests: methodology and implementation at Heshe well site in central Taiwan. Eng. Geol. **198**, 1–15 (2015). https://doi.org/10.1016/j.enggeo.2015.09.006

Yoon, J.S., Hakimhashemi, A., Zang, A., Zimmermann, G.: Particle based discrete element modeling of hydraulic stimulation of geothermal reservoirs, induced seismicity and fault zone deformation. Tunnel Undergr. Space **23**, 493–505 (2013). https://doi.org/10.7474/tus.2013.23.6.493

Yoon, J.S., Zang, A., Stephansson, O.: Numerical investigation on optimized stimulation of intact and naturally fractured deep geothermal reservoirs using hydro-mechanical coupled discrete particles joints model. Geothermics **52**, 165–184 (2014). https://doi.org/10.1016/j.geothermics.2014.01.009

Yoon, J.S., Zang, A., Stephansson, O., Hofmann, H., Zimmermann, G.: Discrete element modelling of hydraulic fracture propagation and dynamic interaction with natural fractures in hard rock. Procedia Eng. **191**, 1023–1031 (2017). https://doi.org/10.1016/j.proeng.2017.05.275

Author Index

A. Tapase et al. (Eds.): GeoChina 2021, SUCI, pp. 185–186, 2021.
https://doi.org/10.1007/978-3-030-79644-0